Farewell to the Internal Clock

Gunter Klein

Farewell to the Internal Clock

A contribution in the field of chronobiology

 Springer

Gunter Klein

ISBN 978-1-4899-9195-9   ISBN 978-0-387-69358-3 (eBook)

Library of Congress Cataloging-in-Publication Data

Klein, Gunter
    Farewell to the internal clock / Gunter Klein, Peter Becker
    p. cm.
    Includes bibliographical references and index.
    ISBN 0-387-40315-9 (alk. paper)
    1. Chronobiology I. Becker, Peter, 1942- . II. Title.
    [DNLM: 1. Chronobiology. QT 167 K64f 2005]
    QP84.6.K57 2005
    612'.022–dc22   2004058969

Printed on acid-free paper.

9 8 7 6 5 4 3 2 1

springer.com

For Maria and Maike

Foreword

Nearly everything making up what we call the "environment" of a plant has an influence on the way it grows. Sunlight, temperature, moisture contents of soil and atmosphere and vibrations are all obvious examples of environmental components, and transient variations in their amount or intensity lead the plant to manifest more or less immediate responses. Small changes in carbon dioxide level in the atmosphere can even have effects, but these take a longer time to be registered – at least those that are visible, albeit at the microscopic level.

Plants meet the challenges of the environment by means of acclimation. In this respect, plants are notable for the plasticity of their development. However, where morphological or physiological plasticity is no longer an option, the responses would be by means of adaptations as a result of genetic selection or genetic "assimilation" (Waddington 1957). Thus, a feature that was once a facultative transient response to an environmental perturbation becomes a constitutive characteristic of plant structure or function. It is in this way that the environment continually molds the way in which plants develop, and also defines the areas upon planet Earth where they will thrive.

Within the "inner space" of the Earth, there are two environmental influences of unvarying intensity (at least within the time frame of geological epochs) and from which nothing can escape. One of these is the gravitational force that attracts two bodies. The second influence is a force imparted by the Moon as it orbits around the Earth. An Earthly gravitational constant of 1G (= 6.670 m 10^{11} N m^2 kg^2) is usually taken for granted except during experiments specifically performed to assess the impact of gravity upon biological development or behavior. Gravity is a uniform background presence during development; it has clearly played a role in shaping the course of plant and animal evolution, and biological constructions are now in harmony with the force that gravity imposes (Barlow 1999). However, even on Earth, not all parts of its land mass present the same gravitational acceleration and, incredible as it may seem, different gravitational domains have been proposed to account for geographically based differences in human development (Kochemasov 1998). In addition, plants and animals have elaborated behavioral responses in the form of gravitropism and gravimorphism (in plants) and gravitaxis and gravikinesis (in animals and unicellular organisms) in order to counteract the pull of gravity. The bases for these responses are not fully understood, but in the case of gravikinesis, the phenomenon seems to rely on the sensation of small redistributions of mass within the selfsame cells that perform these movements (e.g. ref). It is possible that such an autosensation of mass also applies to some aspects of gravitropism and gravimorphoses.

In contrast to Earthly gravitational attraction, tidal forces due to the Moon are not constant but display periodic variations due the rotations of both these celestial bodies. The twice-daily high and low marine tides are evidence of this, though it should be remarked that lunar tidal forces are, of course, present and show similar high and low variations at every point on the Earth's surface, not just at those occupied by seas and oceans. For example, the buildings of the demised World Trade Center in New York are said to have shown a 14-in. uplift in response to the passage of the Moon overhead, and a 2-in. inclination toward the Moon as it set on the horizon (Palmer

2002); and the Large Electron-Positron collider of CERN (Geneva, Switzerland), located in the Swiss Alps at a depth of 100 m and having a circumference of 26.7 km, experiences a few millimeters change in this latter parameter as result of tidal force (Assmann et al. 1994). Thus, it seems unlikely that living systems should have escaped the influence of lunar tidal forces during their development and evolution. On the contrary, just as in the case of the 1G gravitational attraction of Earth, organisms may well have integrated responses to lunar forces in a way that now manifests in their organismal function. In fact, the arguments for tidal influences on behavior of marine and other organisms have been discussed for a number of years (Brown et al. 1970).

The effects of tidal forces on marine organisms are believed to be mediated by the rhythmic increase and decrease of the mass of water that surrounds them. However, such arguments have often not been supported by any real knowledge of the sensitivity of organisms to the pressure changes that were putatively being imposed upon them by their tidal environment. Now that it is known that unicellular organisms exhibit gravikinetic swimming patterns based on a response due to the difference in mass of their own cytoplasm and that of the surrounding medium (Lebert and Häder 1996)), it is evident that sensitivity to the pressure differential at the outer cell membrane – in this case, due to the greater intracellular pressure – can be extremely acute. So, what if it was found that the putative tidal effects were not necessarily a result a lunar force being amplified by the external environment but that organisms could perceive directly a force exerted by the pull of the Moon as it passed overhead in the sky? One consequence would be that organisms all over the surface of the Earth, not just in tidal aqueous environments, would have to be regarded as not only subject to periodic pulls and relaxations of the lunar force but also potentially responsive to these forces. This is exactly what the author of the present little book, the late Dr. Gunter Klein, is proposing. His idea, if extrapolated to its fullest extent, is that plants can perceive directly the pull of the Moon as it passes overhead and that certain plants actually manifest a visible response to this pull in the form of leaf movements.

The daily leaf movements of bean plants (especially species of Phaseolus and Canavalia, for example) have long been the subject of experimental study; and the fact that the leaf movements of undisturbed plants show a rhythm that continues over many days has contributed greatly to the science of chronobiology. In this area, the well, known observations of Professor Erwin Bünning (1906–1990) in Germany helped give rise to the idea of endogenous, or autonomic, rhythms in plants and animals. His ideas were presented in his book, *Die Physiologische Uhr* (Bünning 1958), which was soon translated and published in English as *The Physiological Clock* (Bünning 1964). The autonomic movements appear to be a manifestation of time keeping by an internal clock. The big question in this area has been the nature of the entraining environmental stimuli for the internal clock: that is, what stimuli reset the clock, and how is this accomplished within the plant. Many persuasive experiments have shown that light can bring about significant resetting of the clock, and from here, it was a short step to conclude that daylight or day length was somehow responsible for regulating the natural rhythms of leaf movement. Although it is true that in an experimental situation many stimuli, of which light is one of the most effective, can initiate a rhythmic leaf response, it is possible, nevertheless, that there could be some other, more fundamental, environmental factor at work and that autonomous leaf rhythms have nothing to do with day length. Dr. Klein argues that rhythms of leaf movement are fundamentally under the influence of a physical force that is a function of the position of the Moon. It is the same force that regulates the marine tides: it is called the "lunar tidal force" here.

One of the puzzling features about leaf movements is that their minimal period does not exactly correspond to 24 h, as might be expected on the strictly diurnal basis of the Earth's movement relative to the Sun. Rather, the period is lengthened by about 0.8 h each successive day. A second curious finding is that the period of the rhythms is insensitive to temperature. This is rather remarkable for a so-called "physiological clock" because metabolic processes in plants are generally temperature sensitive (Sutcliffe 1977) – usually they have temperature coefficients (Q10) of approximately 1.6 when estimated experimentally at temperatures in the range 15–25°C. A system that was directly linked both to the sidereal day and to metabolism would be likely to exhibit the two curious features mentioned earlier. However, if the system were regulated directly by a lunar tidal force rather than to day length and/or metabolism, then these features would be exactly as expected. It is such a means of timekeeping, as seen in plant leaf movements, that Dr. Klein is suggesting. However, this proposal does not deny those results that show that other means of entrainment are possible. Clearly, light effects, for example, do regulate the clock. However, underlying them, as a type of default timekeeping process, is the ability to perceive and respond to lunar forces. The default lunar-based response is, perhaps, often masked by more dominant entraining stimuli, and only when these are absent can the default system make itsself apparent.

How does one go about gaining evidence of a lunar basis for plant leaf movements and then of persuading a naturally sceptical scientific community that the lunar hypothesis merits further attention? Many scientific enterprises that seek to establish "cause and effect" relationships rely on some type of statistical correlation for their support. In the case of the lunar hypothesis, this is relatively easy to obtain. Given that the plants are maintained in a suitable environment free from other entraining influences, then what is required is simply a careful record of leaf movements and a knowledge of the lunar tidal force and its variation during the time when leaf movement observations were made. Regrettably, Dr. Klein died before he could assemble his results in this direction into a form suitable for inclusion within the manuscript of the present book. He did, however, describe the method by which he made his observations (see Appendix), and he also left a series of graphs of leaf movements, prepared on known dates, for individual plants. With the help of Professor Emile Klingelé (Swiss Federal Institute of Technology, Zürich, Switzerland, lunar tidal forces have been computed for the location, and on the same dates, Dr. Klein's leaf movement observations were made. Examples of these data have accordingly been inserted Dr. Klein's text, as he surely would have done, in order to support his hypothesis.

However, even before Dr. Klein's experimental observations became available, I had become sufficiently intrigued by his hypothesis to do some analysis on my own account. For this, use was made of the pioneering results of Dr. Anthonia Kleinhoonte, obtained between 1928 and 1931, when she worked at the Botanical Laboratory in Delft, The Netherlands (Kleinhoonte 1932). Kleinhoonte observed individual leaf movements of the jack-bean plant, *Canavalia ensiformis*, and published tracings of these from individual plants on specified dates.

Anthonia Kleinhoonte was an experimentalist and was concerned with the conditions (mainly short pulses of light, or dark periods interspersed with light periods, and other variations on this theme) that shifted the periodicity of the autonomous leaf movements. Nevertheless, among the many published curves of leaf movements it is possible to find cases where the plants appear not to have been subject to experimental environmental influences that would have disturbed the natural rhythm of the movements. Again, Professor Klingelé kindly computed the lunar tidal forces for Delft on the days when leaf movements had been observed by Kleinhoonte. Here, there appears to be a remarkable correspondence between the turning points in the course of the tidal force and the turning points in the course of the leaf movements (Figure 1).

The interrelationships of these two features are just as predicted by Dr. Klein's hypothesis in the pages that follow.

From the point of view of the above analyses, it is important to consider the individual leaf movements of individual plants. This is because each plant has its own characteristic pattern of such movements. Within a group of plants, leaf movements are not synchronized, and therefore an average movement estimated for the group would reveal nothing of value. The asynchrony of movements within a group (Figure 2) presents no conceptual problem, however, and can be understood in terms of a lunar entrainment hypothesis. In fact, the asynchrony is just what would be expected because each plant would respond

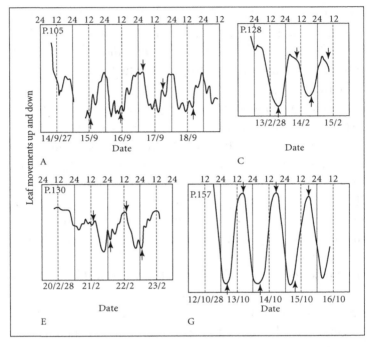

Figures 1A, 1C, 1E, 1G. Tracings of leaf movements of jack-bean, Canavalia ensiformis, plants. These tracings were prepared in Delft, The Netherlands, by Anthonia Kleinhoonte on 14-18/09/1927 (1A), 13-15/02/1928 (1C), 20-23/02/1928 (1E) and 12-16/10/1928 (1G), and were published as Figures 10, 12, 13 and 14, respectively, in Kleinhoonte (1932). The arrowheads indicate times when the corresponding lunar tidal forces for these dates showed turning points (see Figures 1B, 1D, 1F, 1H).

in its own way, in accordance with the different times of the turning points in the continuum of tidal force. As described by Dr. Klein, once a leaf movement – either from "up" to "down", or from "down" to "up" – has been initiated by a tidal turning point, a new movement should not be initiated until the next tidal turning point. If a movement does not commence, however, then the time for the initiation of a new movement is

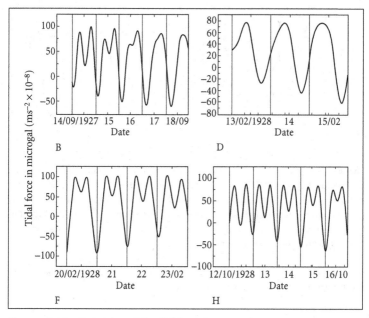

Figures 1B, 1D, 1F, 1H. Plots of the tidal force, in microGals × 10−8 ms-2, computed for the location of Delft (latitude 52° 1' 5", longitude 4° 30' 0", altitude 10 m) on the same dates that Kleinhoonte made her observations of leaf movements. The tidal force turning points indicated in Figures 1A, 1C, 1E and 1G correspond to the troughs and peaks in these plots. Tidal forces were computed by Prof E Klingelé, Swiss Federal Institute of Technology, Zürich, Switzerland.

deferred until the next tidal turning point. Moreover, once a leaf movement has commenced, it might have to reach some natural endpoint before the system is competent to commence a new movement. Thus, the movement itself imposes a lag period before the next new movement can commence.

The only way in which the hypothesis of lunar involvement in leaf movements can be established is if many replicate studies

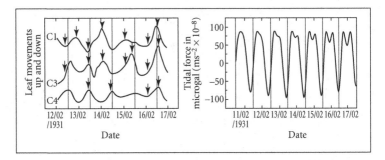

Figure 2A. Tracings of individual leaf movements from three separate plants, C1, C2 and C4, record ed on 12-17/02/1931, in Delft, and published as Figure 17 in Kleinhoonte (1932). For clarity, the original tracings have been smoothed and data for plant C3 omitted. The arrowheads indicate times when the corresponding lunar tidal forces for the dates mentioned showed turning points (see Figure 2B). The tracings also reveal a lack of synchrony of the leaf movements.

Figure 2B. Plots of the lunar tidal force, computed for 12-17/02/1931, which help interpret the data in Fig. 2A. Other details as recorded in the legend for Fig 1B.

are made of leaf movements together with precise computations of the lunar tidal forces. An interesting scenario would be for observations to be made by independent workers in different laboratories, all working simultaneously with plants raised from a common stock of seeds. Lunar forces and turning points in their progress would be different for the different laboratory locations and these might result in different temporal patterns of leaf movements. However, for the hypothesis to remain sustainable, the leaf movements should relate to the different lunar tidal turning points. The next step would be to perform further observations, but under completely different experimental circumstances. For example, just as researchers of gravitropism have availed themselves of microgravity facilities in space to assess the impact of gravity on graviperception (Fitton and

Moore 1996), so similar facilities, especially the International Space Station, where long-term experiments can be carried out, could be used to assess the impact of lunar forces on leaf movements. By escaping from the confines of Earth into a space station where the plants would orbit the Earth more than once a day, a completely new relationship between the Moon and the experimental plants could be set up. An interesting spatial location for an experiment would be at one of the four Langrangian equilibrium points within the Earth–Moon orbiting system. At these points, the centripetal acceleration of a body is perfectly balanced by the gravitational attraction between the Earth and Moon. Hence, there would here be a constant lunar tidal force.

A further set of experimental interventions into leaf movements would inevitably lead to glimpses of the possible endogenous control mechanisms of the movement process and, hence, of the ways in which they might possibly interact with lunar forces. One might hazard that a candidate macromolecule for interactions in the molecular and cellular components of the process is actin, which, in turn, can couple with myosin to thereby form a plant "muscle". Also important would be membrane channels for water and ion flux. Mutations and drugs are known that interfere with actin assembly and thus render leaves immobile or with different sensitivities (and hence provoking different temporal patterns of leaf movements) to the entraining, possibly lunar, influence. Actin, actomyosin, and their interaction with phytochrome might also give clues as to why leaf movements are sensitive to far-red light: far-red light not only abolishes bean leaf movements (Bünning 1973) but also drives the configuration of actin into a form that does not support cell growth (Waller and Nick 1997).

The turning points in the plots of tidal force correspond to moments when the rate of change of the force is zero, or close to zero. That is to say, the force is changing from a "pull" to a "relaxation". At these times, a force is still present, however. It is difficult to envisage how a turning point, per se, could translate into a leaf movement, even with (or without) the assistance of actin and ionic fluxes. Moreover, such lunar forces, even though ever-present, are, according to our conceptions, exceedingly small. In order to be effective, they have to be sensed over and above the background of thermal motion (noise) within the system. The same situation holds for plants in relation to their susception of gravity. Here, mechanisms exist to amplify the graviperceptive signal. In the present case, the question would be whether an analogous amplification system operates in the capture of lunar forces. Although this would be a project for the future, one might guess that possibly all cells of the plant are potentially at the disposal of lunar forces and that a summation of cellular pressures could be focused on certain especially sensitive cells responsible for the leaf movements.

All new ideas meet resistance from supporters of the prevailing culture or scientific paradigm. Nevertheless, if there is a logic to them, new ideas can be tried out, and then their consequences can be either accepted or rejected. For example, Isaac Newton's work on the splitting of white light into its component colors seemed radical at the time; but now, whether light is made of particles or waves, or even whether photons exist until they are "called upon" to come into existence, are problems that puzzle 21st-century physicists (Malin 2001). Scientific progress is a question of extending the horizon of knowledge by the application of increasingly

finer levels of observational resolution. In this way, phenomena that were hidden come into view, and even become the
relative certainties of tomorrow. Moreover, in some uncanny
way, discoveries occur at times that are more or less appro-
riate for them to occur. It is as though the human mind has
already been prepared for them by the "Zeitgeist", or Spirit of
the Age. The impact of the Moon on life on Earth has been
suspected for centuries. Maybe now, at the start of the 21st
century, the time is right for science to "get to the bottom" of
such phenomena. Dr. Gunter Klein's book should help focus
attention on one small area of plant life and enable ways to
be devised that would show whether or not plant biologists
should say farewell to the internal clock – and, presumably,
welcome the era of the lunar clock.

Peter W Barlow DPhil, DSc
Senior Research Fellow
School of Biological Sciences
University of Bristol
Bristol, UK

Foreword References

Assmann R, Blondel A, Dehning B, Grosse-Wiesemann P, Jacobsen R,
    Koutchouk J-P, Miles j; Placidi M, Schmidt R, Wenninger J 1994.
    Energy calibration with resonant depolarization at LEP in 1993,.
    CERN Publication SL/94-61 (Paper presented at 4th Particle Accelerator
    Conference (EPAC 94), London, UK, 27 June-1July 1994).

Barlow PW 1999. Living plant systems: How robust are they in the absence
    of gravity? Advances in Space Research 23 No 12: 1975-1986.

Brown FA, Hastings JW, Palmer JD 1970. The Biological Clock. Two Views.
    Academic Press, New York.

Bünning E 1958. Die Physiologische Uhr. Springer-Verlag, Berlin.

Bünning E 1964. The Physiological Clock. Springer-Verlag, Berlin.

Bünning E 1973. The Physiological Clock. Revised 3rd edition. English Universities Press, London, and Springer-Verlag, New York.

Fitton B, Moore D 1996. National and international space life sciences research programmes 1980 to 1993 – and beyond. In: Moore D, Bie P, Oser H (eds) Biological and Medical Research in Microgravity. Springer-Verlag, Berlin. pp 432-541.

Hemmersbach R, Bräucker R 2002. Gravity-related behaviour in ciliates and flagellates. In: Advances in Space Biology and Medicine, Volume 8, Cell Biology and Biotechnology (ed. A. Cogoli), pp. 59-75. Elsevier, Amsterdam.

Kleinhoonte A 1932. Untersuchungen über die autonomen Bewegungen der Primärblätter von Canavalia ensiformis DC. Jahrbücher für wissenschaftliche Botanik 75, 679-725.

Kochemasov G 1999. Dichotomy of Earth in tectonics and human races. In: Non-traditional Aspects of Geology, VII Scientific Seminar Proceedings, "Harmony in Structure of Earth and Planets", pp. 45-47, Moscow State University, Moscow.

Lebert M, Häder D-P 1996. How Euglena tells up from down. Nature 379, 590.

Malin S 2001. Nature Loves to Hide. Quantum Physics and the Nature of Reality, a Western Perspective. Oxford University Press, Oxford.

Palmer JD 2002. The Living Clock. Oxford University Press, Oxford.

Sutcliffe J 1977. Plants and Temperature. Studies in Biology No. 86. Arnold, London.

Waddington CH 1957. The Strategy of the Genes. Allen and Unwin, London.

Waller F, Nick P 1997. Response of actin microfilaments during phytochrome-controlled growth of maize seedlings. Protoplasma 200, 154-162.

Contents

ARISTOTELIS DIVINI UNIVERSUM...

GALILEO: *The truth is not the child of the authorities, but the child of the times. Our ignorance is unbounded; let us reduce it by a cubic centimeter! What is the use of wanting to be so wise, when we can finally be ever so slightly less ignorant! I have had the unimaginable happiness of getting hold of a new instrument with which one can see a tiny corner of the universe somewhat more clearly, not much. Try it for yourselves.*

THE PHILOSOPHER: *Your Majesty, Ladies and Gentlemen, I only wonder where all this will lead.*

GALILEO: *I would say that as scientists it is not our concern where the truth may lead us.*

THE PHILOSOPHER, *wildly: Mr. Galileo, the truth might lead us anywhere!*

GALILEO: *Your Highness. These nights, all over Italy, telescopes are being turned toward the heavens. The moons of Jupiter will not lower the price of milk. But they have never been seen before, and yet they exist. From this the man on the street draws the conclusion that there could be many more things to see, if he would only open his eyes! You owe him a confirmation of this! It is not the movements of a few distant stars which has made all Italy sit up and take notice, but the knowledge that teachings regarded as unshakable have been called into question, and everyone knows that there are too many such teachings. Gentlemen, let us not defend discredited teachings!*

FEDERZONI: *You, as the teachers, should assist in discrediting them.*

GALILEO: *But the gentlemen really need only look through the instrument!*

(Bertolt Brecht, *Life of Galileo*, Scene 4)

> *There is a great difference between **still** believing in something and believing in it **again**. To **still** believe that the moon influences plants betrays foolishness and superstition, but to believe it **again** shows the influence of philosophy and reflection.*
>
> G.C. Lichtenberg

For many people, the Moon is a familiar acquaintance with mysterious qualities. Depending on how it is illuminated by the Sun, it shines sometimes as a complete disk and sometimes only as a sickle, or else makes itself invisible. The forces emanating from the Moon give rise to high and low tides on the seacoasts. Its course and orbital velocity can be calculated exactly by means of complex mathematical formulas. In 1969, human beings visited the Moon for the first time, and brought back rocks. Frozen water is thought to exist at the poles; perhaps this means that the time is approaching when there will be a human settlement on the Moon. This satellite of the Earth is moving away from us at the rate of 4 cm per year and is as old as our 4.7 billion-year-old world. Today, only the origin of the Moon remains a mystery. Was it catapulted from the Earth by a cosmic super collision, or was it a wandering stray, captured by our planet from the cosmos? Apart from this, the relationship between the Moon and the Earth is considered to be all but completely understood by science.

Nevertheless, in addition, there seems to be another type of knowledge, which receives no attention in the textbooks. The media report on it, and esoteric discussions of it fill whole bookshelves in the bookstores. According to this alternative point of view, the Moon is not simply a sphere of inert material, proceeding impassively on its course.

Rather, it is said to possess mystical powers that can exert an influence over terrestrial life. Such "occult wisdom" – as it styles itself – urgently advises that the correct phase and position of the Moon be chosen for everything from a visit to the doctor to dental care. Gardeners are advised, among other things, to dig up the vegetable garden in the Spring only when the waxing moon appears in the constellation of Leo.

For the uninitiated, the question arises as to whether the influence of the terrestrial satellite is really so far-reaching. Should one follow this type of guidance? Just because of the method of communication, information disseminated by the media gives many people the impression that everything transmitted by this means must necessarily have some truth to it.

"Mystic humbug," say the enlightened natural sciences. There is no evidence for the existence of such supersensory powers of the Moon.

This book is not a contribution to the numerous superstitious beliefs relating to the Moon. Rather, the following will demonstrate, and above all **prove by means of experiments that can be reproduced at any time** that the Moon does, in fact, influence certain biological rhythms, evoking them and determining their timing.

Scientists have hitherto attributed this phenomenon to an "internal clock." According to this hypothesis, the living organism itself has the ability to measure time and to determine how much time has elapsed.

Paradoxically, it appears in the end that the theory formulated by science in this problem area actually rests only on "mystic foundations," whereas the basic view of the esoteric and of widespread folk beliefs, that there is a direct connection between the moon and biological functions, in fact comes closer to the truth.

In order to ensure their reliability, the experiments that provide the proof of this have, in part, already been performed over 400 times. They are described in detail in the Appendix of this book and can be repeated using simple materials. By this means, it should be possible for every reader to gain the necessary conviction, on the basis of his or her own research.

In the beginning was the leaf

Most life processes are carried out in recurring cycles. This is the case for plants, animals, and human beings. The periods of such rhythmic events vary greatly, ranging from brain waves measuring tenths of seconds, to 24-hour sleep/wake cycles, to yearly rhythms. All living beings have close relationships to one another and to their environment. Temporal coordination and adaptation are required in order to ensure existence and continued survival. For example, this applies to the simultaneous reproductive readiness of sexual partners, the flowering of plants, annual winter hibernation, and the punctual migration of birds. The field of chronobiology is committed to investigating these phenomena. Its goal is to address questions concerning the causes and practical significance of such temporal structures. How does an organism know the time of day or the time of the year? Although mankind has been able to observe these regularly recurring phenomena repeatedly, both in himself and in the environment, scientific research in this field began relatively late.

In the year 1729, the French astronomer De Marian observed that during the day, when illuminated by sunlight, the leaflets of a mimosa plant assumed a horizontal position, whereas in the evening, when it became dark, the leaves folded up. The next morning, the incident light again caused the leaves to unfold. These leaf movements were repeated day after day. After De Marian had placed the plant in a light proof cupboard, he ascertained that the leaf oscillations also took place in the absence of the daily light/dark cycle of outdoor light. He noted: "Thus, the mimosa feels the sun without in the least being able to see it" (DeMairan 1729). According to his interpretation, a type of rhythmic motion must be involved that could be generated by the plant itself, in the absence of external influences.

A friend of the astronomer sent a report of these findings to the Communications of the Royal Academy, however, it met with no response. In the field of natural history, this was the era of classifying and cataloging the individual species of plants and animals; at that time, there was scarcely any general scientific interest in special, isolated questions.

Seventy-five years later, the Swiss botanist Augustin de Candolle discovered that the leaf movements of the mimosa also responded to an artificial light source. He placed the plants in artificial light during the night and in darkness during the daytime. After some time, the mimosa adapted itself to these altered conditions, spreading out its leaflets at night when the light was shining, and folding them up again in the daytime, when it was dark. Even when the lamps burned uninterruptedly day and night, the plant moved its leaves as in duty bound. However, the scientist ascertained that during this continuous illumination, the mimosa was not able to imitate the 24-hour solar day exactly; the plant miscalculated the length of the day by a few hours each time (De Candolle 1832).

Charles Darwin likewise investigated the movements of plants. He placed them under various lighting conditions, darkened parts of them, used mechanical means to hinder the nocturnal folding of the leaves, and examined the influence of sounds by having his son play on a horn. He came to the conclusion that movements caused by external factors (e.g., light) are to be differentiated from those generated autonomously (by the plant itself) (Darwin 1896).

The studies were gradually extended to the rhythmic behavior of animals and human beings. The rest and activity periods of rats, mice, and human infants attracted scientific interest; research in the field of chronobiology had begun.

With the establishment of systematic investigations, the bean plant became a standard research object. It was quick and easy to grow, and the leaves displayed a conspicuous oscillatory behavior. The leaves are raised and lowered in an exact 24-hour rhythm, adapted to natural daily lighting conditions. Under the influence of sunlight, the leaves are maintained in an almost horizontal position; however, when darkness falls, they bend down until they are vertical. At the beginning of the day, they are again raised toward the incident light. As the leaves are lowered, the leaf stalks attached to the stem are raised upward, like pointers; they are lowered again as the leaves are raised. This behavior has been referred to as "sleep movements." The raising and lowering movements are effected by means of hinge like articulations between the leaf stalk and the leaf and between the stem and the leaf stalk (Hensel 1993).

The Dutch researcher Kleinhoonte experimented with jack beans. She raised these plants under the unnatural regime of 8 hours of light alternating with 8 hours of darkness – that is to say – a 16-hour day. The bean plants adapted themselves to these conditions and altered their leaf positions accordingly at 8-hour intervals. The researcher then placed the plants under conditions of continuous illumination or uninterrupted darkness. As if it were inborn, under these constant lighting conditions, the bean plants moved their leaves in approximately the rhythm of a 24-hour day (Kleinhoonte 1928).

Then, around 60 years ago, the German biologist Erwin Bünning systematized the investigation of rhythms. He, likewise, worked mainly with bean plants (particularly with the scarlet runner bean). In the view of this biologist, research using plants could make just as valuable a contribution to the general elucidation of biological rhythms as research using animals. He was primarily interested in rhythmic events taking place during the course of a single day (Bünning 1977).

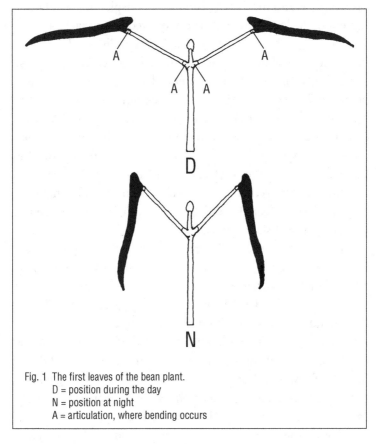

Fig. 1 The first leaves of the bean plant.
D = position during the day
N = position at night
A = articulation, where bending occurs

His findings can be summarized as follows. He determined that in the case of plants, animals, and human beings in the natural environment, rhythmic oscillations of biological processes are evoked by the regular alteration of various external influences. The foremost of these influences is the light/dark cycle of daylight. However, marine plants and animals at the coasts orient themselves primarily according to the varying water levels of high and low tides. In the case of human beings, social contacts can also be a decisive factor. High and low temperatures and variations in atmospheric humidity seldom act as rhythm generators. The environmental factors that act on organisms and direct the periodically fluctuating biological processes have been termed "time givers" (Ehlers, Frank, and Kupfer 1988).

If the decisive external time giver is eliminated, as when, for example, a mimosa or a bean plant is placed in constant artificial illumination or continuous darkness, for the majority of living beings the rhythms continue to be carried on. Bünning concluded from this that the various organisms are able to generate such recurring processes without any help from the environment and are also able to determine for themselves the intervals at which the rhythmically occurring life processes are repeated. Thus, like the bean plant, other organisms must also possess an "internal clock," which sets the pace independently of any information from the natural environment.

However, strangely enough, the length of a day as measured by the internal chronometer does not correspond exactly to a 24-hour time span. The self-generated rhythms deviate by as much as several hours from the length of a solar day. Because this correspondence is only approximate, the daily cycles of the "internal clock" are referred to as "circadian" rhythms (from the Latin "circa" = approximate; "dies" = day). For instance, according to Bünning, in continuous light bean plants sometimes required an average of 28 hours for the execution of their leaf oscillations; the movements were delayed considerably compared to movement behavior exhibited outdoors under daily fluctuating lighting conditions (Bünning 1977).

Bünning could only conjecture as to where this "internal clock" might be located and how it functions. All that could be said for certain was that the operation of the "internal clock" must be affected by biochemical means. However, chemical processes proceed at different speeds, depending on the temperature. For instance, it is well known that detergent acts more quickly in warm water than in cold water. It had been found, however, that the precision of the "internal clock" was practically unaffected by differences in external temperature. This was inconsistent with the laws of chemistry. In order to justify the "internal clock" theory, another capability was therefore attributed to the organism. It must have the ability to eliminate the influence of fluctuations in temperature on the operation of the clock. However, how this could be effected remained obscure.

One now had – for example, in the case of the bean plant – two clocks, so to speak. There was an "external" clock, which, in the natural environment, was guided by the daily light/dark cycle of sunlight. Only in the absence of the time giver from the environment did an "internal" clock become evident, reacting automatically at regular intervals. However, the temporal cues generated by the two clocks did not coincide. The "light clock" ticked in an exact 24-hour rhythm, whereas the other was not even accurate to the nearest hour.

Bünning dealt with this the problem by viewing the "internal clock" as the actual rhythm generator, which, in turn, receives a corrective influence from the external time giver, the day/night cycle. This time giver shortens or lengthens the duration of the inexact cycles specified by the internal chronometer, so as to reconcile them with the external lighting conditions of the

24-hour day. This is comparable to a conductor who uses a baton to direct an orchestra to play a piece of music faster or slower. According to this explanation, the "internal clock" is thus synchronized by an external factor. However, the details of how this is brought about could not be explained (Bünning 1977).

As Bünning's theory of the "internal clock" became internationally known, he did not earn only praise for his work. Many colleagues considered the hypothesis of the existence of such an internal chronometer to be bizarre, mysterious, extremely fanciful, or ridiculous. Until the middle of this century, it was still fashionable in scientific circles to make fun of the so-called endogenous (internal) rhythms.

Chronobiology today

The science of chronobiology is no longer limited to the field of biology. Its effects can be seen in medicine, pharmacology, psychology, pedagogy and ecology. Thus, specialized fields have been established such as chronomedicine, with further subdivisions, and chronopharmacology (Gutenbrunner, Hildebrandt and Moog 1991).

The findings concerning the "internal clock" are considered to be conclusive. Among other things, bean plants have been subjected to the most varying conditions. The plants have been brought to the vicinity of the South Pole, transported as flying objects, and studied in mines. These measures are designed to verify that no previously unknown external influences (such as cosmic radiation or magnetic forces) could be involved in stimulating the "internal clock." Because the bean plants continue to move their leaves with constant regularity despite all of the diverse experimental conditions, it is considered to be scientifically proven that these rhythms could be generated only by the organism itself. The "internal clock" has an assured place in chronobiology. Bünning's theory has been adopted almost unqualified. Environmental factors such as light alterations and tides determine when the orders given by the organism's internal chronometer will be carried out. This external control is extremely necessary, in order to adapt the inexact internal rhythms to natural conditions. Otherwise, the bean plant might, for example, end up in a situation where it is standing with lowered leaves in full sunlight, being unable to make optimal use of the light necessary for its development because of the imprecise timing of its self-generated rhythms.

"Internal clocks" are part of the basic equipment of life and are found in all plants, animals, and human beings. According to the almost unanimous conviction of scientists, they are innate and function according to the same laws in all living beings.

The rhythmic reactions of the internal chronometer can occur at intervals that are shorter than a day (ultradian), approximately corresponding to a day (circadian) or significantly longer than 24 hours (infradian). These constitute further conceptual subdivisions. The majority of the "internal clocks" known at the present time generate circadian rhythms. The light/dark cycle of daylight is the most widespread time giver. It adapts the initially imprecise rhythms specified by the internal chronometer to the exact length of a day (Chadwick and Ackrill 1995).

Meanwhile, in innumerable international publications, scientists have been concerned with the phenomenon of rhythm generation. It is assumed that every living being possesses an "internal clock." Only a small selection of typical cases can be presented here by way of example.

The marine alga Gonyaulax, which consists of only a single cell, exhibits three different daily recurring processes. In the middle of the night, the alga shines with the most intensity, giving rise to what is referred to as marine phosphorescence. At midday, photosynthesis, the use of sunlight to synthesize energy-rich substances, reaches its maximum. Shortly before dusk, cell division occurs. If the alga is kept in the laboratory in constant, continuous weak light, these processes continue to function. There are therefore three separate circadian rhythms that can be generated by the alga itself. Thus, there must be room for the "internal clock" within a single cell (von der Heyde, Wilkens, and Rensing 1992).

Many kinds of animal are especially active at particular times of day. For instance, golden hamsters and cockroaches regularly begin to run at certain times. Rats and lizards also have varied activity patterns during the day. This behavior continues to be exhibited in experimental situations where there is no natural fluctuation of light, with the movement impulse being guided exclusively by a circadian "internal clock" (Aschoff, Daan, and Groos 1982).

A constant daily temporal rhythm likewise underlies the emergence of the fruit fly *Drosophila* from its pupa. In the natural environment, the duration of this rhythm is determined by the daylight/darkness time giver however, in the absence of fluctuations of light, the rhythm continues to function freely, directed by the organism itself (Dowse, Hall, and Ringo 1987).

In the case of the bread mold *Neurospora crassa,* the "internal clock" brings about the creation of a pattern. This fungus, which accidentally made its way into scientific laboratories in 1924, has since then, as a research subject, contributed to revolutionary discoveries in the fields of genetics and molecular biology. There is even a separate periodical devoted to it. During its growth phase, under natural lighting conditions *Neurospora crassa* produces its asexual spores once a day, at a particular time. Even in the absence of the alternation of day and night, under conditions of constant illumination or continuous darkness, this cycle of development continues to function. However, the original 24-hour rhythm is not maintained exactly. It now takes place in a circadian time period, which corresponds only approximately to the length of a day (Feldman and Hoyle 1973).

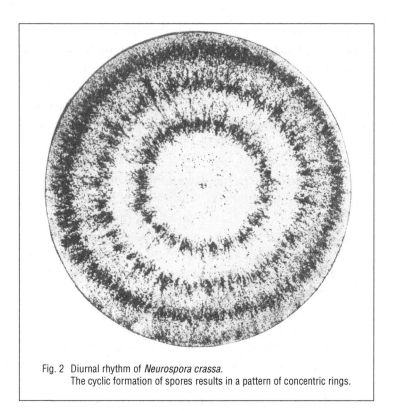

Fig. 2 Diurnal rhythm of *Neurospora crassa*.
The cyclic formation of spores results in a pattern of concentric rings.

Like many other animals, birds also display an annual rhythm. In this case, the changing proportion of daylight in a 24-hour day acts as the external time giver. From the beginning of the year, the period of daylight increases each day to a maximum in the summer, and then gradually decreases again until the end of the year. The birds thus obtain signals from their en-

Fig. 2 is taken from: RENSING, L.: Der molekulare Mechanismus der circadianen Uhr, in: Biologie in unserer Zeit 25 (1995) S. 101

vironment for timely mating, nesting and departure in the fall on the migration to their planned winter quarters in the south (Farner and Lewis 1971).

Garden warblers, for example, were kept in cages under conditions of constant illumination and unvarying temperature. Using only their "internal clock," the animals roughly maintained their natural annual rhythm for more than 3 years. In their migration at the beginning of autumn, garden warblers first fly southwest from their summer quarters in Europe to northwest Africa, crossing the Straits of Gibraltar. From there, in late autumn they continue their journey to their winter quarters in central or south Africa (Gwinner 1996).

This second stage of their journey requires the birds to make a radical change in the direction of their flight. In order to reach their final destination, they must now fly toward the southeast. Warblers held in captivity under constant environmental conditions displayed a time-shifted directional behavior in the cage, which corresponded to the change of course of an unhindered fall migration. An "internal clock" thus seems to provide the birds with assistance in navigation, enabling them to determine the beginning, duration, and temporal progression of their migration to the south.

Human beings, in particular, have now become the focus of chronobiological research. It has been found that numerous bodily functions and behavior patterns occur in daily cycles, which are guided by the daylight/darkness time giver (Lemmer 1996).

For example, waking and sleeping are related to the daily fluctuations of light. If these signals suddenly deviate from the pattern of the accustomed 24-hour day, significant disruptions in the functioning of the human body can result.

Particularly in the case of long-distance flights to the east or west, flight personnel and travelers are constantly subjected to changing local times. The fluctuation of daylight no longer takes place in a regular 24-hour rhythm. This causes difficulties with respect to the adjustment of the body rhythms, with the associated unpleasant symptoms referred to as "jet lag." The reason is the delay of the "internal clock." It is difficult for it to adjust immediately to the shift in the light/dark cycle of daylight brought about by the change in time zones. Those affected often display lowered productivity, problems with sleeping, and, in extreme cases, even acute depression, for days or weeks (Preston 1973).

The first traveler to suffer from this was the American aviation pioneer Wiley Post. In 1931, he planned to fly around the world with his monoplane. Two days after the beginning of the trip, Post had been thrown off balance by the resulting changes in daylight to such an extent that on his flight to Berlin, he had to return to Hanover. The reason was that he had forgotten to refuel his plane (Orlock 1995).

Comparable symptoms appear in the case of night-shift workers who are subjected to an unnatural rhythm of light fluctuations (Roden et al. 1993).

In winter, around 20 % of North Americans succumb to what is referred to as "winter depression," because of the weak outdoor light. Those affected are depressed, find it difficult to get up in the morning, and tend to gain weight. In the spring, these symptoms gradually disappear. Scientists have found that bright light can shift the biological rhythms. For instance, a small spotlight has been developed that can be worn on the head. At the push of a button, this shines into one's eyes, giving rise to at least a temporary alertness (Lewy, Sack, and Singer 1985).

In one experiment, the fly *Drosophila* was subjected to artificially altered light/dark cycles corresponding to a 21-hour or a 27-hour day. The animals experiencing these cycles died perceptibly sooner than members of the same species in the natural environment (Bargiello and Young 1984).

However, it is not only the sleep/wake behavior of human beings that fluctuates at regular intervals each day. Almost all regulated bodily functions and a large number of faculties and perceptions likewise exhibit a daily cycle. For instance, body temperature alters in the course of the day. At 4 o'clock in the morning, it is at its lower limit of approximately 36.1° Celsius. Toward midday it climbs to around 36.8° and at 8 p.m. reaches a maximum of over 37° afterward sinking again to its minimum.

Corresponding oscillations also occur in blood pressure, saliva production, insulin and cortisone levels, and kidney function (Chan, Folk, and Huston 1968).

The immune system has a lower resistance at night than in the daytime. Mood, attentiveness, memory and sensitivity to pain also fluctuate daily.

The cells of the body divide every 24 hours, with tumor cells dividing at a different time of day than healthy cells.

In addition, in the human body, alcohol tolerance is dependent on the time of consumption. Experiments with mice have shown that alcohol injections at one time of day result only in the animals becoming "tipsy," whereas at another time of day, they have lethal effects (Reinberg and Smolensky 1983).

Experiments were required in order to address the question of whether these daily rhythms occur in human beings even in the absence of the natural light/darkness signal giver. All other regularly recurring events, such as sounds, social contacts, and punctual mealtimes also had to be eliminated in order to ensure that such events could not be used as orientation cues for determining the passage of time. The test subjects therefore had to be placed in total isolation.

In 1989, an Italian volunteer, Stefania Follini, moved into a transparent plastic container buried 10 meters deep in the desert of New Mexico. She was subjected to continuous electrical illumination and was under constant observation. Video cameras recorded all sounds, as well as her movements; in addition, she had to carry out numerous medical tests. The experiment was concluded after 130 days.

All subsequent experiments with research subjects in "time isolation bunkers" also showed beyond any doubt that the sleep/wake behavior, general state of mind, and bodily functions maintain their characteristic rhythmic oscillations even in the absence of external environmental influences.

Only the period of the rhythms deviates from the 24-hour day; at 25 hours, it corresponds only approximately to the length of a day. Thus, the self-generated cycles are again circadian (Moore-Ede, Sulzman, and Fuller 1982).

It had now been established that from bean plants to humans, for every living being the operation of the "internal clock" is programmed into the organism. This internal timekeeper specifies the actual rhythms, which are then shifted to the "right time" of the 24-hour day by a time giver from the environment – usually the natural light/dark cycle of daylight. If there is a breakdown or interruption of this process, the smooth operation of the interdependent bodily functions is impaired.

The benefits for medicine were obvious. Within the general area referred to as "chronomedicine," various fields of study were developed in order to apply the findings of chronobiology advantageously in the field of medicine.

Industrial medicine is concerned with the relationship between light fluctuations and shift work.

A chronotherapy clinic in Texas has the research aim of ascertaining the correct timing for the treatment of illness. With regard to cancer therapy, experiments with mice suffering from leukemia had already shown that sensitivity to radiation differed, depending on the time of day. Animals that were irradiated at a certain time all died. When the same dose was administered to a second group at a different time of day, all of the animals survived (Reinberg and Smolensky 1983).

It was known that body cells divide every 24 hours. However, the time of greatest activity with regard to cell division varies from tissue to tissue. Thus, in the treatment of cancer in human beings, it is important to determine the time of the least cell division activity by healthy cells in the affected organs, in order to minimize the undesirable side effects of the treatment (Hrushesky 1993).

The optimal effect of many other medical treatments is likewise dependent on the time of day that they are administered. For instance, blood pressure is relatively low during the night and is highest in the early morning (Deanfield and Sellier 1994). In the light of these fluctuations, it would, therefore, be appropriate to reduce the medication of high blood pressure patients at night and to increase it in the early morning when the rise in pressure occurs. The standard prescription of "one tablet every morning, noon, and night" could be altered so as to be more appropriate and effective, depending on the clinical picture. However, the findings of chronobiology are only beginning to be taken into consideration with regard to medical treatment methods.

According to one chronobiologist, it would be "a blessing for medicine" if it were finally possible to understand completely the mechanics of the "internal clock," so as to be able to influence its temporal cues.

Scientists at McGill University in Montreal recently found that a certain species of nematode lived five times longer than usual when the time interval between cell divisions was lengthened.

The Moon and chronobiology

The fluctuation of light between day and night is not the only decisive external time giver and pacemaker for the regulation of internal rhythms.

In their natural environment, many marine plants and animals orient themselves exclusively to the alternation of the high and low tides caused by the Moon. In western Europe and North America, these exhibit a rhythmic period of approximately half a day; however, in other places, they may occur only once a day, for example in southeast Asia or in the Gulf of Mexico.

Species living right at the coast must adapt their rhythms to the changing level of the ocean, in order to search for food at the right time in the area uncovered by the retreating water at low tide or in order to be able to use the tidal currents for dissemination or reproduction. For around 600 species of marine plants and animals studied so far, it has been scientifically confirmed that the tides act as the time giver for the regulation of a great variety of biological rhythms (Palmer 1995).

These marine organisms have been brought into the laboratory to eliminate the influence of external time givers. Even in the absence of the rise and fall of the water level, numerous species continue to exhibit the rhythms previously displayed in their natural habitat.

There is a marine sand flea that lives in the sand of the Californian coast. This amphipod leaves its place of residence as soon as the beach is flooded, and buries itself again at low tide. Researchers have placed animals from the coast in an aquarium and observed them under conditions of constant temperature, continuous illumination, and unvarying water level. Even under these constant conditions, it was found that at recurring intervals, the sand fleas left their hiding places in the sand and began to swim in the water, afterward disappearing again into the sand. The intervals between emergence and disappearance corresponded to the tidal rhythms at the seashore. The scientists concluded that the animals must have an "internal clock" that "ticked" in time with high and low tide (Palmer 1995).

Oysters open their shells longer when the tide is coming in than when it is going out, in order to utilize water currents from the ocean as carriers of fresh supplies of food. These mollusks also exhibit the same behavior in stationary water in the laboratory.

The difference between the water level at high and low tides is not always the same. At intervals of approximately 7.4 days, there are extremes in the extent of the change. Exceptionally powerful tides (spring tides) alternate with weak tides (neap tides). This phenomenon is connected with the angle of the Sun, the Moon, and the Earth with respect to one another, because the Sun also exerts a gravitational attraction and thus either reinforces or weakens the gravitational pull of the Moon. A few life, forms have specialized so as to regulate the duration of their rhythms exclusively according to the onset of a spring tide or a neap tide, each of which occurs approximately every 2 weeks.

Grunion, which grow up to 15 cm long, inhabit the Pacific Ocean off the coast of California. During the spring and summer, large schools of these fish appear at regular intervals near the coast. The timing of this large-scale phenomenon is precisely synchronized. Three to four nights after each new and full moon (the phases of the Moon which give rise to the spring tides), during the first 2 hours after high tide, grunions allow themselves to be carried up onto the beach by incoming waves.

With strong swimming movements, they resist being dragged back by the retreating water. The females lay eggs, which within 30 seconds are fertilized by the males. Then, the fish, without resistance, are swept back into the ocean by the next wave. The eggs incubate in the beach sand during the subsequent period of normal tide levels, when the waves do not wash over the sand. Precisely at the return of the next spring tide, the young fish hatch out of the eggs and are swept into the ocean by the waves, which now reach far up the beach. This reproductive cycle proceeds regularly, in the rhythm of the spring tides (Palmer 1995).

In 1954, two American scientists published the results of several years' study of the breeding behavior of sooty terns inhabiting a tropical island in the Atlantic Ocean. These birds do not breed in a yearly rhythm; rather, the interval between breeding times amounts to barely 10 months. Each year, the terns thus lay their eggs some 2 months earlier than in the preceding year.

It was observed that this breeding rhythm corresponded very exactly to the length of 10 lunar months, the period of time required by the terrestrial satellite to orbit the Earth 10 times. This led to the question of whether the birds regulate their behavior according to the Moon and have the ability to count to 10.

Subsequently, a German biologist kept African stonechats under constant conditions in the laboratory for several years, in order to study their breeding behavior. In their natural habitat in Kenya, these birds breed in a 12-month yearly rhythm. Under conditions of constant temperature, unvarying food supply, and regular light fluctuations, the birds bred at intervals averaging 9.3 months. Because under these constant conditions the stonechats could not observe the course of the Moon or other changing external influences, the researcher concluded that the breeding time was specified solely by an "internal annual clock." In the natural habitat, during the course of a year, the increasing and decreasing day length would act as a time giver from the environment that would, so to speak, slow down the "internal clock," which was going much too fast, to regulate it to 12 months. Thus, only the synchronization of the clock by means of corrective external influences enables the birds to adapt to the natural yearly cycle (Gwinner 1996).

Various rhythms having periods corresponding to a lunar cycle are also found in human beings. As early as the beginning of the nineteenth century, the Czech naturalist Purkinje found that in twilight, the human eye perceives individual colors with differing acuteness. During the transition from day to night, red shades seem to fade first, whereas blue tones can be recognized the longest as the darkness increases. Further studies have shown that this difference in the color sensitivity of the eye can also be reversed. On days with a full moon, as dusk falls, the red shades can be recognized for a longer time, whereas the blue tones fade first. This phenomenon occurs even when the test subject does not set eyes on the full moon.

Scientists have also repeatedly confirmed that in human beings, the excretion of uric acid diminishes at the time of each new and full moon. This occurs completely independently of whether or not the Moon itself is actually seen (Berndt et al. 1995).

In 1977, American physicians reported on a physically healthy 28-year-old man who had been blind from birth. The man suffered from severe sleep disorders, which greatly impaired his work and leisure activities. The patient was placed in isolation, in a clinic, for almost 4 weeks. The data gathered during this time showed that the rhythms of the various bodily functions were not adapted to a 24-hour day, but had a period of 24.85 hours. His sleep behavior displayed a day/night rhythm of 24.9 hours. Each of these time periods corresponded to a lunar day – in other words, the time required by the earth to revolve on its axis until the next reappearance of the moon (Miles, Raynal, and Wilson 1977).

In addition, the doctors noticed that the beginning of sleep always coincided with a low tide at the local coast. Evidently, the blind man could not adapt exactly to a 24-hour day on the basis of external stimuli. His bodily functions were regulated solely by his imprecise, unsynchronized "internal clock," which generated only circadian rhythms. His internal rhythms were thus continuously shifted with respect to the demands placed on him by daily events at work and during his leisure time.

In the view of chronobiologists, individual biological rhythms are always regulated in a similar manner. In the absence of external influences, the organism itself with its "internal clock" determines when and at what intervals the recurring events, such as leaf movements, the opening of mussel shells, or waking and sleeping, are to occur. However, the temporal cues provided are inexact. In contrast, the environment demands a punctual adaptation to relevant events and resets the internal chronometer accordingly.

According to the prevailing opinion of chronobiologists, this fundamental assumption, considered to be well proven, necessarily means that the role of the Moon is exclusively limited to causing the high and low tides. These tides brought about by the Moon are consequently only an external corrective time giver for the "internal clock," enabling marine organisms to harmonize the rhythms generated by the organism itself with water level conditions at the relevant coast. Any type of more direct, far-reaching influence that might be exerted by the terrestrial satellite is disputed by science.

Insofar as the "internal clock" "strikes" at time intervals that correspond to the various lunar rhythms, this is attributed to adaptations to lunar cycles important for survival (such as the tides, spring tides, and the full and new moon) that have arisen in the course of evolution. During the evolution of the organism, these rhythms have been "programmed and internalized" and are thus no longer connected with actual external lunar events currently occurring in the sky.

In contrast to this view, folk wisdom persists in adhering to a belief in a more far-reaching influence of the Moon on plants, animals, and human beings. There are reports of sleepwalkers, magically attracted by the full moon, who rise out of bed and climb onto the roof. In addition, many people are convinced that they can get no rest on nights when the Moon is full. Others are allegedly prone to excessive alcohol consumption at this time, which has led to the terrestrial satellite being referred to by the colorful colloquial expression "drunkard's sun."

Peasant proverbs connect the phases of the Moon with a change in the weather, and the most severe earthquakes in southern California are said to have occurred at the time of the full moon (Ertel 1996).

An influence on the human psyche is likewise attributed to this heavenly body. The American psychiatrist Lieber collected crime statistics for 15 years and is convinced that the full moon can trigger acts of violence. In love, it is said to make one "mad." For instance, Buddha is credited with saying that he had to lock up the monasteries when the Moon was full, so as to prevent the monks from fulfilling their desires (Lieber 1997).

"Complete nonsense," retorts science. There is no evidence for such "lunar power." The statistics presented are contradictory. To some extent, psychologists account for the popular belief in the influence of the moon by means of the phenomenon of the "self-fulfilling prophecy." According to this explanation, a feeling, a mood, or a mode of behavior often does, in fact, arise at the time that its occurrence is expected. The thought of being certain to suffer from disturbed sleep at the time of the next full moon is enough in itself to cause such anxiety that, with the appearance of the bright disk in the heavens, as expected, one cannot sleep a wink.

There is also the Moon as "assistant in every emergency." Here, on the one hand, the focus is on whether the terrestrial satellite is in a waxing or a waning phase. On the other hand, the current position of the Moon in one of the 12 signs of the zodiac in the starry sky is said to be decisive. Special lunar calendars permit those seeking advice to keep themselves informed of the phase and position of the Moon each day. The recommendations of the authors, which are based on intuitive feelings, alleged perceptions, and traditions handed down from ancient times, are often mutually contradictory.

Nevertheless, the fundamental principle underlying these beliefs is: The waxing moon assists with everything that is to grow and prosper, whereas the waning moon assists with everything that is to be diminished (Spieß 1994). The rules given for gardening, medical treatment, beauty care, home, and leisure are generally based on the "theory of analogy." This prescientific way of thinking forms the basis of the belief that like must give rise to like. For instance, because of its color, red wine, as the "blood of the earth," is thought to be particularly effective as a means of fortifying the blood. According to this principle, hair that has been cut should grow back more quickly when the Moon is waxing. Vermin, weeds, and stains are to be combated when the Moon is waning, so that their presence will diminish like that of the moonlight.

Skeptical science views this way of living "in time with the Moon" as being "behind the times," because it is at variance with all present-day knowledge concerning the weak material effects that moonlight can have on any biological function. In addition, as far as the position of the Moon relative to the signs of the zodiac is concerned, the already familiar arguments against "astrological superstitions" apply.

Location and mode of operation of the "internal clock"

After the discovery of the "internal clock," chronobiologists faced the question of where in the organism the clock is located and how it functions. To date, these problems have been solved only partially or not at all.

The first question to be addressed is the mode of operation of the external time giver, which synchronizes the "internal clock" so that it is adjusted exactly to the 24-hour day. In beans and other plants, the alternation between sunlight and darkness is perceived by a receptor in the plant cell. In the cell is a protein called phytochrome. This has the property of becoming active and reacting in the presence of incident light. In darkness, it loses all effect. During its active phase in the presence of light, phytochrome causes numerous biochemical processes in the plant to be set in motion. These ultimately ensure that the daily leaf movements of the bean occur "on time," in synchrony with the light/dark cycle (Edmunds 1988).

Where in the bean plant the internal chronometer that actually generates the leaf movements might be located remains as much a mystery as the mechanism of the clockwork.

In this regard, somewhat more exact information is available for human beings and other mammals. An "internal clock" that evokes their daily rhythms is presumed to exist in the region of the base of the midbrain. There, directly above the intersection of the two optic nerves, is the so-called "suprachiasmatic nucleus." In this structure, which is not even as big as the head of a pin, the autonomous "central internal clock" is said to be located. Many brain and body functions are regulated from here (Moore 1983).

However, as has been shown by experiments with animals and human beings in isolation chambers, this internal chronometer is also not synchronized precisely with the 24-hour day. Once again, the rhythms generated must then be adapted exactly to the natural light/dark cycle of the environment. In human beings and other mammals, this daily fluctuation of light is first perceived by means of the eyes. The retina then transmits the relevant light intensity information by means of the optic nerves, and the pineal gland at the top of the midbrain is, in, turn stimulated to produce the hormone melatonin. This hormone is then distributed to nearby parts of the brain and to the bloodstream (Armstrong 1989).

The melatonin level in the blood undergoes marked diurnal fluctuations. The daytime value is five times the nighttime amount. This can be attributed to the fact that the pineal gland adjusts its hormone production in accordance with signals concerning the degree of illumination as perceived through the eyes. Like the bean plants, our bodies also receive chemical information indicating current lighting conditions in the environment. It is assumed that the imprecise "internal clock" is reconciled exactly with the rhythm of daylight by means of the fluctuations in the level of melatonin, which are controlled by the time giver – the light/dark cycle of outdoor light.

Medical research has shown that melatonin produced by the pineal gland in excessive amounts or over too long a period can lead to winter depression or to the travel ailment referred to as jet lag. Therefore, in treating these symptoms, melatonin is used as a medication, or light therapy is used to mediate the production of this hormone, in order to reset the "internal clock" (Lewy 1983).

The German biologist Bünning hypothesized that the mechanism of the "internal clock" could be based on a vibrating system, an oscillator (Bünning 1977). In such a system, two different states (e.g., of a chemical nature) alternating continuously at regular, equal intervals would have to be generated in certain cells. These regularly recurring changes of state could then be used as units for the measurement of time. This is comparable to a pendulum clock, where the alternating swings of the pendulum, taking place at equal intervals, can serve as a basis for measuring time.

The modern quartz watch is likewise based on this principle, but in this case, it is crystal vibrations that determine the rhythm. Even the second of our chronology is no longer the $86,400^{th}$ of the mean length of a solar day; this would be too imprecise. Independently of the orbit of the Earth around the Sun, a second corresponds to exactly 9,192,631,770 vibrations of a cesium atom. Such oscillating atomic clocks achieve a precision of operation that varies by only 1 second in a million years.

The search for such a vibratory system in an organism was at first unsuccessful. One was dependent on theories alone. Chronobiologists spoke of a breakthrough when a recurring process with an approximately daily rhythm was observed in the cells of a few life-forms.

In the nucleus of each cell are the genes, the storehouses of encoded hereditary information. In order to act, the genes must first send their information out of the cell nucleus into the interior of the cell. For this purpose, they make use of a messenger material, referred to as "messenger ribonucleic acid" (mRNA). This obtains all of the necessary information from the gene and then makes its way out of the nucleus into the cell. In the course of the further transmission of information, the mRNA then forms a protein. The messenger material mRNA is being constantly produced by the gene; as the amount of mRNA increases, the amount of protein generated by the mRNA also increases. For individual genes, it has been found that this protein then migrates back into the cell nucleus, where it combines with the gene. By this means, the gene is prevented from generating more mRNA, and as a result, the amount of mRNA is gradually reduced. In this way, the protein switches off the production of its own manufacturer, and after some time, it likewise becomes depleted. In the absence of the protein, the gene can then "breathe again"; it once more produces mRNA for the transmission of information, and the cycle starts again from the beginning (Rosato, Piccin, and Kyriacou 1997).

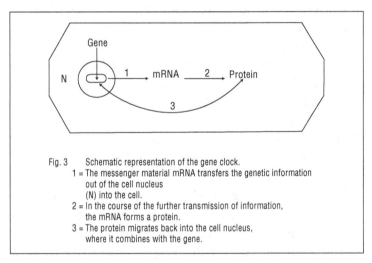

Fig. 3   Schematic representation of the gene clock.
1 = The messenger material mRNA transfers the genetic information
     out of the cell nucleus
     (N) into the cell.
2 = In the course of the further transmission of information,
     the mRNA forms a protein.
3 = The protein migrates back into the cell nucleus,
     where it combines with the gene.

This "stop-and-go procedure" in the course of genetic activity, together with the associated increase and decrease in the amounts of mRNA and protein, takes place continuously in the individual cell at intervals corresponding approximately to the length of a day. In the smallest possible space, an "internal clock" had been discovered that "ticks" in a circadian rhythm. Such a clock gene, measuring time in the above-described manner, has so far been found only in the bread mold *Neurospora,* the fruit fly *Drosophila,* and the mouse (Rensing 1995). However, following this discovery, chronobiologists still faced the decisive question: What, in turn, controls the regular circadian operation of this molecular "internal mini-clock"? It might be expected that directly after the first formation of protein by the mRNA, a "short circuit" would occur, with the protein immediately penetrating the cell nucleus and deactivating the gene. However, this, in fact, occurs only after a certain waiting period, so that the cycle of starting and stopping of the reactions stretches over the length of approximately a whole day. Although this gene clock certainly provides material for further hypotheses and demonstrates that temporal cues are produced in the cell, nevertheless, even after the discovery of these rhythmic processes, the details of how time is determined in the cell still cannot be explained in a sufficiently satisfactory manner. Likewise, no regulatory system has been found that could enable the biochemical gene clock to "tick" at a constant speed, unaffected by the external temperature.

No discoveries have yet been made concerning the location and mode of operation of the "internal clock" that controls weekly, monthly, or yearly rhythms.

In summary, it can be concluded that the mystery has not yet been solved.

The mistake concerning the "internal clock" of the bean

Everything actually started with the bean. Its leaf oscillations, which continue even in the absence of the light/dark cycle of natural daylight, are still the standard textbook example used in schools and universities as proof of the existence of an "internal clock."

Conceptions concerning the nature and mode of operation of this time measurement affected by the plant itself remain hazy. It must, therefore, be considered permissible to return once more to the decisive fundamental question of whether the organism can, in fact, execute its leaf movements under constant lighting conditions by itself, "without outside help." Or are there forces operating from the outside, which have up to now been overlooked or simply left out of consideration?

A relevant experiment could be based on the following theoretical consideration: If a hitherto unknown factor in the environment exerts a central control over the leaf movement rhythms of bean plants under constant lighting conditions, plants of the same type should all receive and respond to such an external signal at the same time. The result would be that the movement reactions of the different plants would occur in complete harmony with one another. This is similar to the principle of a radio clock, for which a transmitter synchronizes all of the clocks in its area that are ready to receive the signal.

In order to test this, some 10 to 12 bean plants of the same species and commercial variety may be raised under conditions of natural outdoor light. As soon as all the plants are following the same rhythm, stretching their fully grown leaves toward the sunlight in the morning and lowering them again when darkness falls in the evening, the experiment can begin. The bean plants are to be moved one by one, at irregular intervals, and placed under conditions of constant continuous illumination.

During the next few days, it appears that the plants execute their leaf oscillations completely independently of one another. The leaves of some bean plants are horizontal and the leaves of others are vertical; at times, some leaves are descending, whereas others, on the contrary, are ascending – no trace of togetherness. Evidently, each plant goes its own "rhythmic way."

The Dutch scientist Kleinhoonte made these observations as early as the first third of this century (Kleinhoonte 1928). She concluded that no force acting from the outside could be involved in controlling the leaf movements, otherwise the rhythms of all the bean plants would be synchronized. There was now nothing to hinder the development of the idea of an "internal clock."

However, if the experiment is not broken off after these initial observations have been made, a surprising discovery results. It is first necessary to record the timing of the oscillatory behavior separately for each individual plant, noting the time of the turning point of a leaf movement to the nearest minute. The turning point is the moment at which the bean plant begins either to lower its leaves from the horizontal position or to raise them from the vertical position.

The results immediately reveal that it is apparently only at certain times that the reversal movements occur. With regard to these movements, all of the bean plants remain inactive for hours at a time. Then, suddenly, a larger or smaller proportion of the plants exhibits a simultaneous change in the direction of movement of their leaves. As if "with one accord," some bean plants begin to lower and others to raise their leaves then. In contrast, the remaining plants appear completely unaffected and continue their leaf movements unchanged in the original direction. After this, there is another pause lasting several hours, in which none of the test plants execute a reversal movement.

Then, the performance is repeated. Once again, a few of the bean plants alter the direction of their leaf movements at the same moment, while the remaining plants continue their movements unchanged for the next few hours.

This rhythmically occurring "partial group behavior" continues to be exhibited and could be represented schematically as follows: Part of the group reverses direction – pause (no plants reverse direction) – part of the group reverses direction – pause (no plants reverse direction) – and so forth.

The length of the pauses between turning points, when not one of the bean plants executes a reversal of movement direction, is always at least 6.2 hours. However, the value can also be 12.4 or even 18.6 hours (plus or minus a few minutes each time).

There are, thus, certain fixed times at which turning points may occur. However, these are not employed simultaneously by all of the test plants; they are used in stages, by a portion of the plants each time. An at first inexplicable order now emerges from the initially seemingly unrelated confusion of the rhythms. In fact, each bean plant moves on its own, timing its turning points by means of its own "internal clock," with the result that changes in the direction of oscillation of the leaves of individual plants continually occur at irregular intervals.

It follows that there must be some type of system underlying the rhythms of the leaf movements. If one assumes that the plants in a group do not have the ability to "come to an agreement" with one another in timing their movements, then only an external factor could be responsible for inducing a change in movement direction at fixed intervals.

However, strangely enough, such a signal from the environment seems to reach only some of the bean plants, compelling them to react. The remaining plants at first remain unaffected, but sometime later, they also "take their turn," again after the passage of fixed periods of time. When a single bean plant is put to the test for several days, the situation is scarcely more comprehensible. Under constant illumination, it appears that the interval between two points at which the direction of movement is reversed (corresponding to half an oscillation) is, as a rule, either 12.4 or 18.6 hours. For example, 12.4 hours can elapse from the time the leaves are lowered until they are raised again, and another 18.6 hours can pass before they are lowered once more. Thus, a complete oscillation does not always consist of two half-oscillations of equal length. Occasionally, the time difference between the two turning points of a half-oscillation is only 6.2 hours. The duration of a half-oscillation alters seemingly at will, with the bean plant selecting a time interval equal in length to one of the periods mentioned. This means that the duration of the individual leaf oscillations can constantly vary.

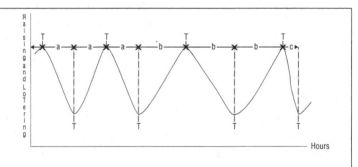

Fig. 4   Various possible temporal configurations of the leaf movement cycle, based on half cycles varying in length.
T = turning point
a = 12.4 hours; b = 18.6 hours; c = 6.2 hours

In attempting to explain an unfamiliar phenomenon, one should first apply what is already known. Only when all such means of explanation fail should unknown causes be assumed. With respect to the leaf movement experiments, this implies searching for known forces that could be responsible for the behavior exhibited by the bean plants in the tests.

An abstract inspection of the numbers representing the varying time spans between the turning points of the leaf movements of individual plants or between the reactions of groups of plants reveals that the length of the intervals is almost always 6.2, 12.4, or 18.6 hours. No special consideration is necessary in order to ascertain that this is a number series based on a factor of 6.2 and its multiples.

Only one known force affects the Earth in a 6.2-hour rhythm. This is the tidal force, which generates the high and low tides. As is well known, it is caused in part by the Moon, which reaches its highest point with respect to the observer once every 24 hours, 50 minutes, 28 seconds on the average, after somewhat more than one full rotation of the Earth (Kumm 1992).

The "lunar day" is, therefore, somewhat longer than the solar day of 24 hours, because the Moon travels in the same direction as the Earth's rotation. An observer on our planet thus overtakes the Moon again only after a little more than one complete rotation of the Earth has occurred.

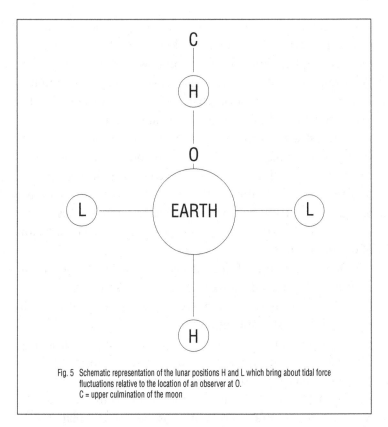

Fig. 5 Schematic representation of the lunar positions H and L which bring about tidal force fluctuations relative to the location of an observer at O.
C = upper culmination of the moon

During the course of such a lunar day, the influence of the tidal force at a given location fluctuates, increasing and decreasing every 6.2 hours, as can be seen from the regular alternation of high and low tides at our coasts. Whenever the Moon reaches its highest point in the south (or culmination) with respect to a given place on the Earth's surface, the effects of the tidal force at that place will be at their strongest. The tidal force will again exert its strongest influence at the same place when, after somewhat more than half a rotation of the Earth, the Moon reaches its highest point at the spot directly opposite that place, on the other side of the terrestrial globe. In Figure 5, these two positions of the Moon are indicated by H (= high tide), for an observer at location O. The tidal force exerts the least influence when the Moon reaches either of the positions indicated by L (= low tide) relative to the location O. (The physical principles underlying the origin of the tides will not be described here in detail; a discussion of these can be found in the relevant technical literature.)

As stated earlier, the movement of the terrestrial satellite from position H to L, or L to H, takes an average of 6.2 hours. The time required for the Moon to travel between two positions designated by same symbol is, therefore, 12.4 hours. The lunar day, from the time that the Moon reaches its highest point with respect to a given place to the next time this occurs, lasts a total of 24.8 hours.

This number series corresponds to that found for the length of the pauses, or the intervals between turning points, in the above-described experiments. Whether coincidental or not, and although neglected by all previous theories in the field of chronobiology, this similarity is to be thoroughly investigated.

Before beginning the decisive experiments, it is necessary to calculate the exact time at which the Moon reaches positions H and L relative to the test location, because these are the positions associated with fluctuations of the tidal force. (The method of calculation is described in the Appendix II.)

Several bean plants, which have previously displayed their usual oscillatory behavior in sunlight and darkness, are then placed, simultaneously or at intervals, under conditions of uninterrupted illumination. Once the plants begin their regular leaf movements, the exact time of all the turning points, the reversal from a rising to a falling, or a falling to a rising motion, is recorded for each bean plant for several days. The time is recorded to the nearest minute.

A comparison of these values with the previously ascertained times of the H and L positions of the Moon yields the following results.

For all of the plants, in principle, each reversal movement following a completed leaf oscillation begins only when the Moon occupies one of the H or L positions relative to the test location. It should be noted that each reversal of the tidal force associated with an H or L position of the terrestrial satellite can evoke the beginning of either a downward or an upward oscillation of the leaves. For a reversal of leaf movement, the sole decisive factor is that the Moon must be located at one of the four positions. The behavior of the bean plants thus differs from the action of the high and low tides at the coast, where the relevant H positions of the Moon are always associated with a high water level and the L positions are always associated with a low water level.

In addition, the bean plants do not make use of every fluctuation of the tidal force in order to alter the direction of their leaf movements. For example, if a plant has executed a movement reversal when the Moon is at one of the H positions, it will usually react again only after 12.4 hours, at the time of the next H position, disregarding the intervening L position. It is equally possible that after its turning point, the plant will disregard both the next L and the next H position of the satellite, "skipping over" them and only after 18.6 hours, at the time of the subsequent L position, executing the change of direction of the previously begun leaf oscillation. This gives rise to the frequently observed variations in the length of the periods of the individual rhythms. Sometimes, the reversal movement occurs at the time of the immediately following fluctuation of the tidal force, after 6.2 hours.

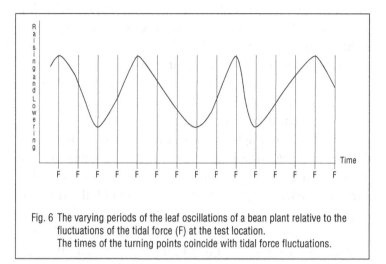

Fig. 6 The varying periods of the leaf oscillations of a bean plant relative to the fluctuations of the tidal force (F) at the test location.
The times of the turning points coincide with tidal force fluctuations.

In summary, the following can be concluded. Under constant lighting conditions, each turning point, and hence each change of direction of the leaf movements of the bean plants, regularly occurs at the same time as an astronomical fluctuation of the tidal force, which is brought about by the changing position of the Moon relative to the location of the plants. Consequently, the varying period lengths of the individual oscillatory rhythms always correspond to a multiple of the interval between the fluctuations of the tidal force at a particular location.

Thus, it is not an "internal clock" that determines the rhythm for plants placed under conditions of continuous illumination or darkness.

From the very beginning of chronobiological research, it was observed that for numerous life-forms the circadian rhythms that appeared under constant lighting conditions always corresponded approximately to the length of a solar day. The American biologist F. A. Brown Jr. consequently hypothesized a connection with the daily fluctuating conditions of atmospheric pressure or the changes in gravitation brought about by the rotation of the Earth (Brown 1960). Brown's hypothesis contrasts sharply with the increasingly firmly entrenched theory of the existence of an "internal clock."

In order to shed light on the controversial question of whether circadian rhythms might, in fact, be attributable to some environmental influence connected with the rotation of the Earth, in 1960, scientists from the University of California, Los Angeles traveled to the South Pole. At a research station some 800 meters from the geographical pole they conducted experiments with hamsters, the mold fungus *Neurospora,* the fruit fly *Drosophila,* cockroaches, and garden beans. Among other things, these organisms were placed on turntables that were rotated in a manner exactly counter to the rotation of the Earth. This resulted in the same effect as if the Earth had ceased to rotate. The test subjects were thus maintained in a constant position with respect to the stars. Regardless of this, the bean plants raised and lowered their leaves under conditions of continuous darkness in the research laboratory (Hammner 1962). The investigators concluded from this that external factors based on the daily rotation of the Earth exerted no influence on the basic mechanism of the "internal clock." In the view of most, but not all, scientists, the hypothesis that time measurement in the plants might be controlled by an external influence was thereby refuted.

At first glance, this might seem to imply that in the case of the bean plants, priority is given to adapting the period of the leaf movements to the light/dark cycle of sunlight. In the absence of this, the plants, so to speak, configure the length of the diurnal rhythms with the aid of fluctuations of the tidal force. Finally, if these lunar signals are also no longer available, the plants activate their "internal clock," as the experiment at the South Pole appears to indicate. This would be a completely new theory.

In order to clarify this issue further, it is first necessary to consider which factors give rise to the tidal force. The tidal force arises solely from the interplay between the gravitational and centrifugal forces that result from the rotation of the Earth and the Moon, and also the Sun and the Earth, about a common center of gravity. This is not to be confused with the additional factor of the revolution of the Earth about its own axis. This rotation of our planet about its axis makes no contribution to the actual origin of the tidal force, but only brings about the daily 6.2-hour average time shift of the fluctuations of the tidal force.

If our planet were to cease to rotate, this would not eliminate any of the factors that give rise to the tidal forces and influence their changing intensity, because even relative to a stationary Earth, the Moon would constantly alter its position each day, continuing to travel, moving closer to or further away from us, assuming a different position with respect to the Sun, and, above all, altering its angular distance from the equator (declination). The daily varying declination is sufficient in itself to cause a corresponding change in the water level at the sea coasts, even without the simultaneous rotation of the Earth about its axis.

Thus, in the "South Pole experiments," only one environmental factor was eliminated. This factor primarily determines the time interval of the changing intensity of the tidal forces with respect to the Earth, but does not represent the sole cause of their existence on our planet. For this reason, the results obtained from these experiments cannot be used to rule out the influence on the test subjects of "lunar gravitation," the magnitude of which continues to vary. With regard to the leaf movements of the bean plants, there is, thus, no demonstrable proof of the existence of an "internal clock" that "starts up" and regulates the rhythms in the absence of the fluctuating lunar forces.

Unfortunately, the article published by the researchers concerning the biological clock at the South Pole does not include data on the length of the periods of the leaf oscillations. However, the duration of the movement cycles in the experiments could be expected to vary somewhat from that of the raising and lowering movements that would be generated under the additional influence of the Earth's rotation.

Another noteworthy aspect of the experimental results is based on the records of the daily activity patterns of hamsters and the rhythms of emergence of fruit flies at the South Pole. Although the individual groups of test subjects had no contact with one another and were placed in conditions of uninterrupted darkness on turntables undergoing a great variety of different types of rotation, almost all of the animals in these various circumstances executed their daily behavior patterns at the same time, as with one accord. This synchronization of the individual "internal clocks" also suggests the operation of an external influence.

All of the results obtained so far thus permit only one conclusion. In the absence of fluctuations of daylight, only the "wandering" Moon brings the leaves into rhythmic oscillation. The bean plant does not possess its own chronometer, located in the organism itself. The "clock" ascribed to it by chronobiology is not "internal"; rather, it operates by "remote control."

Thus, rather than referring to the leaf oscillation rhythms occurring in the absence of the light/dark cycle of sunlight in terms of their duration, as "approximately daily = circadian," it seems logical to choose a term characterizing the cause of the movement. Hence, in the following discussion, the term "lunagent rhythms" (from the Latin "luna" = moon; "agere" = do, bring about) will be used.

However, as indicated by the experimental results, under constant lighting conditions, the individual bean plant does not make use of each of the tidal force fluctuations, occurring every 6.2 hours, as a signal for the initiation of a reversal movement. Following a turning point, the plant generally omits the next, or the next two opportunities "offered by the Moon" to alter the direction of its leaf movements, before reacting again.

This behavior can be explained in terms of a familiar phenomenon. If a movement-triggering stimulus (in this case, the fluctuation of the tidal force) is received by the plant cells, they become excited by this incoming signal and a reaction is initiated. During the time that the movement is being executed, the excitation of the cells is checked, and in this state, they become insensitive to further signals. The so-called "refractory period" sets in. Only after the "command" already given has been executed completely does the state of the cells return to normal; the plant is then ready to receive new stimuli.

Because the raising and lowering of the leaves under continuous illumination occurs relatively slowly and at varying speeds, at the time of the next, or the next two fluctuations of the tidal force, the bean plant is usually still in the midst of executing a movement. Thus, the refractory period persists and prevents the processing of the next signal, which often occurs too soon for the plant to make use of it. Why the time period between reversal movements does not remain constant for an individual plant, but fluctuates as a rule between 12.4 and 18.6 hours, is not clear.

The average length of time taken for a complete oscillatory movement (raising and lowering of the leaves) as controlled by the "internal clock" is variously reported in the literature. At present, this "circadian" average value is most commonly estimated as being 27 to 28 hours (Chadwick and Ackrill 1995). At first glance, this seems inconsistent with the experimentally obtained results, which are always, and according to the conclusions derived must always be, evenly divisible by a factor of 6.2 (the time interval between tidal force fluctuations). However, the experiments have shown that a complete cycle can be composed of two half-cycles differing in length. For example, if the leaf movements of a bean plant occur during the following time periods: downward for 12.4 hours – upward for 18.6 hours – downward for 12.4 hours – upward for 12.4 hours, the resulting length of two complete cycles is 55.8 hours, which means that the average time for a single cycle is 27.9 hours. Occasionally, short half-cycles of 6.2 hours occur, so that each experiment can result in a different average value. Thus, according to the laws of mathematics, the average values, unlike the individual values, would not necessarily be divisible by the factor 6.2.

The duration of the "lunagent" leaf oscillations of the bean plant is, therefore, not rigidly fixed, but it fluctuates constantly in the above-described manner. This might also have contributed to the fact that the synchrony of the cycles and their dependence on the position of the Moon were not recognized earlier.

The fluctuation of the duration of leaf oscillations also explains the "partial group behavior" and the subsequent "obligatory pause" exhibited by the bean plants in the experiment described initially. In a group of several plants, each individual has its own rhythm. This means that, at a given time, only a few of the plants have finished raising or lowering their leaves and are ready to execute a direction reversal by the time of the next tidal force fluctuation. Meanwhile, the rest of the bean plants have not yet completed the execution of their leaf movements. Their cells are in a refractory state and are, therefore, prevented from reacting to the pending fluctuation of the tidal force. Thus, these plants must wait for a later signal, as soon as they have ended their movements.

Hence, the bean plants hence have the opportunity of changing the direction of their leaf movements every 6.2 hours; however, at a given time, only a portion of the plants can take advantage of this opportunity. If the number of test plants is large enough, according to the laws of probability, every 6.2 hours, or at least every 12.4 hours, a few of the plants in the group will meet the conditions necessary to execute a movement reversal. Except at these times, all of the plants pause, and none reverse their movement direction.

A superficial consideration of the facts might lead to the conclusion that the powerful, and in the truest sense of the word, "earth-shaking" tidal forces can unquestionably also induce the oscillations of bean leaves. These forces not only displace the waters of the oceans but they also literally "tug" at the Earth, the surface of which, together with the Eiffel Tower and the Cologne cathedral, is raised by as much as half a meter and then lowered again. This movement cannot be seen, however, because all of the reference points are displaced by the same amount.

The thermal energy resulting from this deformation of the Earth is estimated to be at least 40 billion watts. Because humans and other living beings consist predominantly of water, these facts have induced a large number of "moon believers" to adopt a "theory of biological tides," This is based on the idea that a force powerful enough to move the waters of the oceans must certainly be able to influence the small amounts of water in a living organism.

However, such a theory is not supported by the law of gravity discovered by Newton more than 300 years ago. This states that all objects attract one another with a force that is directly proportional to the product of their masses, and inversely proportional to the square of the distance between them. Consequently, the effect of the gravitational attraction of the Moon is dependent on the mass of the object that is affected; the smaller the mass of the object, the weaker the effect of the gravitational force (Schäfer 1988). In the Lake of Constance, the tides are only just measurable, whereas in a swimming pool or a bathtub, they might just as well not exist. In addition, the water in an organism is stored in tiny cells. Thus, based on these considerations, the theory of miniature high and low tides in the organism would have to be rejected.

However, in associations of living cells, the situation appears to be not quite so simple. Scientists at the Swiss Institute for Technology in Zurich recently discovered that, just as in the oceans, high and low tides also exist in trees. Exactly in time with the lunar tidal rhythm, the tree trunks swell minimally when the terrestrial satellite is at a high tide position and shrink again when it reaches a low tide position – not enough to be perceived with the naked eye, but by a measurable amount. However, this occurs only as long as the cells are alive. According to the researchers, the increased gravitational attraction of the Moon holds the water in the living cells, whereas the decreased gravitational force allows the water to flow out again. Thus, the reaction to the tidal forces of water contained in cells seems to differ from that of loose drops of water.

Nevertheless, this discovery cannot be transferred directly to the influence of the Moon on the oscillatory behavior of bean plants under conditions of constant illumination. Whereas the high and low tides at the seacoasts are linked to particular lunar positions, the raising of the bean leaves is by no means always associated with an H position of the moon, nor is the lowering of the leaves necessarily associated with an L position. The bean plants can raise or lower their leaves when the terrestrial satellite is at either an H or an L position. For the initiation of a reversal movement, it is only necessary for the Moon to occupy one of the positions resulting in a tidal force fluctuation; it does not matter which type of position it is.

The influence of the tidal forces on the bean plants must, therefore, operate in a different manner. By means of investigations using modern techniques, the internal mechanism of the "circadian" leaf movements of the bean plant has been fairly well explained. A hingelike articulation is found between the leaf and the leaf stalk. On the upper and lower sides of this articulation are specialized tissues, which are linked to one another. If all of the cells have the same internal pressure and are filled with water so that they are "equally fat," the leaves remain in the horizontal position. Before the leaves are lowered, the tissue at the top becomes excessively full of water and increases in volume. At the same time, the amount of water in the lower cell region, located opposite, is reduced. The tissue at the bottom slackens completely and no longer offers sufficient resistance to pressure from the cell network above, with the result that the leaves bend down to the vertical position (Simons 1994).

However, before this displacement of water can occur, some preparatory measures are necessary. The swelling of the top of the articulation through the accumulation of water is attributable to a prior shifting of positively and negatively charged atoms, called ions, between the upper and lower tissues. There is a change in the concentration of certain ions in the upper tissue as compared to the lower tissue. Ion pumps first transport positively charged hydrogen atoms out of the cells of the upper tissue, thereby increasing the electrical potential between the interior and exterior of the cells. A greater number of potassium and chloride ions then enters the cells, and this, in accordance with the law of osmosis, causes water to flow in after them. The internal pressure of the cells increases, the upper tissue expands, and the movement of the articulation results (Hensel 1993).

In the case of high and low tides at the seacoasts, the tidal forces act directly on the masses of water, which are thus raised or lowered. In the case of the leaf articulation mechanism of the bean plant, in light of the above discussion, it can be seen that the influence of the tidal forces must be restricted to an effect on the steps preceding the change of water concentration in the tissues of the cells.

The question thus arises as to whether the fluctuating pull of the Moon could in any way influence the redistribution of the ions. Because of the small mass of such atoms, according to the law of gravity discovered by Newton, it would be expected that, in principle, the tidal forces of the Moon would be unable to have any effect.

In 1992, researchers at the CERN Institute for High Energy Particle Physics in Geneva made a curious discovery. The scientists accelerated atomic particles along a path many kilometers in length, until they reached velocities close to the speed of light and eventually collided with other particles of matter. Upon collision, the particles involved burst apart into subatomic constituents, detectable only by means of their effects. The whole facility is located 100 meters beneath the surface of the Earth, in order to eliminate any external influences. The particles are accelerated around a ring-shaped path 26.6 kilometers long that has been hewn out of the rock.

One difficulty with the experimental procedure was that the particles easily became magnetized. In order to eliminate this unwanted side effect, a high-frequency field was installed. However, once this was done, the scientists were surprised to find that the resonance frequency fluctuated on a daily basis. After lengthy investigations, it was found that the length of the path periodically increased and decreased, by 2 millimeters each time. The researchers then determined that this was unmistakably attributable to the effect of the Moon, which altered the length of the path to be traveled by the particles at intervals corresponding to the astronomical tidal rhythms (Gutzwiller 1996).

It was thus apparent that, in principle, the fluctuating tidal forces are able to exert an influence on elementary particles that are in motion. It is true that the effects seem to be extremely slight. However, the crucial point is that such a reaction can occur.

Newton's law of gravity cannot by itself explain the interaction of gravitational forces between individual elementary particles. For instance, light or radio waves originating from distant stars and having a mass of zero are deflected from their course by the gravitational attraction of the Sun. This is due not to the mass of the particles, but to the energy of the light and radio waves. Because under laboratory conditions the gravitational forces of elementary particles are exceedingly weak and experimental investigations are practically unfeasible, scientists have so far been unable to develop a universally valid gravitational theory that could apply to the realm of the smallest particles; much remains undiscovered and unexplained. Thus, in light of the present level of knowledge in the field of physics, it might be very possible that the tidal forces could exert an influence on elementary particles.

It is to be assumed that the energy required for the rhythmic leaf movement reactions of the bean plant is produced and made available by the plant itself. The influence of the Moon would thus be limited solely to a "timer function," which serves only to trigger the flow of energy. It is conceivable that the continuously increasing and decreasing gravitational pull of the terrestrial satellite might affect the cellular "ion environment" in a manner not yet understood. Possibly a change in the intensity of the force could switch ion pumps on or off, or in some other way permit the redistribution of the charged atoms.

The discovery of the "lunagent" rhythms also makes superfluous the hitherto unsuccessful search for the mechanism which allows the plants to maintain their leaf movement reactions at a constant speed, uninfluenced by the external temperature (assuming, of course, that this lies within the range that will support life). The Moon always sends its signals at constant intervals, "rain or shine."

The tidal forces are omnipresent; they penetrate every obstacle and cannot be switched off. Consequently, the bean plants react even in mines or submarines. Because the plants display their leaf movement rhythms in the absence of the light/dark cycle of daylight in the greatest variety of locations, this has been considered to be evidence for the existence of an "internal clock." It was believed that every possible environmental factor that might be involved had been eliminated; the "lunar forces" were apparently overlooked or excluded from the outset as being scientifically inconceivable.

The question of why the bean plants execute the raising and lowering of their leaves, movements that require the expenditure of energy, attracted the attention of Darwin. His conclusion is still accepted today. In the lowered state, the leaves "hug each other," reducing heat radiation and thus preventing excessive heat loss. In the daytime, however, it is important for the survival of the plant for as much light as possible to be captured by means of a horizontal leaf position (Darwin 1896).

Plants orient themselves by means of gravity. The roots tend toward the center of the Earth, and the shoot above ground grows in the opposite direction. This growth form also makes possible the optimal use of the available incident sunlight falling from above, the radiant energy of which is required so that it can be transformed into chemical energy. If the light should happen to fall only on one side of the plant, the latter alters its gravity-oriented position and grows in a curve, toward the light source. The requirement for light takes precedence over the alignment with respect to gravity. There are thus two complementary external "direction givers" (gravity and light) responsible for the growth behavior of the plant. Where these compete with one another, the "light" factor has priority.

A similar "basic idea" of evolution is exhibited in the movements of bean leaves. On the one hand, there is the raising and lowering of the leaves brought about by the daily light/dark cycle. This allows the plant to make optimal use of the incident light. On the other hand, there are the "lunagent" rhythms, which come into play only in the absence of the light-oriented oscillations. Once again, there are two environmental influences that operate according to an order of precedence.

In the field of chronobiology, the concept of synchronization was developed in order to explain these phenomena. According to this concept, the rhythms generated by the "internal clock" have a period that does not correspond exactly to a time span of 24 hours; they are adapted to the exact length of a day by means of the fluctuations of external light. But how exactly is this supposed to take place?

It thus seems more logical to proceed on the assumption of two separate "rhythm givers" (light and tidal force fluctuations), the operation of which is somewhat comparable to the competing influences of gravity and sunlight in the case of plant growth. Occasional experiments seem to point to a certain interdependence of the two systems, without, however, allowing anything definite to be said about a possible interaction.

Does the "lunar clock" also tick for animals and human beings?

It was the leaf movement rhythms of the bean plants that laid the foundation for the hypothesis of the existence of an "internal clock." The question as to whether the same "clockwork" underlies all of the self-generated rhythms that have so far been discovered in various living beings remains unanswered. Only the similarity of the fundamental principles has been confirmed. The occurrence of "lunagent oscillations" in the bean plant is not enough in itself to permit the conclusion that the "lunar clock" also marks time for other living beings. In order to gain a better understanding of this problem, it seems appropriate to sift and evaluate the available scientific "clues and evidence" used to substantiate the existence of an "internal clock."

1. According to the view of chronobiologists, the "internal clock" reveals itself by means of its operation. Its presence can be detected only in laboratory experiments. In such experiments, the influence of all periodically fluctuating environmental factors that could act as external pacemakers must be eliminated. If, under such constant laboratory conditions, the rhythms continue to appear, it is concluded that they could only be generated by the organism itself.

However, as indicated earlier, it is not possible to eliminate the fluctuations of the tidal force. No laboratory door can prevent their effects. A "moon-regulated clock," as such, can function anywhere and can receive the external signals that control it in every research laboratory. So far, all experiments carried out under so-called "constant conditions" have continued to be affected by the constantly fluctuating pull of the Moon. Consequently, such experiments cannot be used to refute the existence of additional "lunagent rhythms."

Biologists have placed the mold fungus *Neurospora* in a space laboratory orbiting the Earth, where for 1 week the fungus clearly exhibited its diurnal rhythmic growth behavior. Researchers were convinced that in this experiment, all geophysical external time givers had been eliminated, and they viewed the results as further proof of the existence of an "internal clock" (Waterman 1990).

It is true that in a space laboratory orbiting the Earth approximately every 90 minutes, the gravitational pull of our planet is reduced to a few thousandths of its value at the surface of the Earth. However, in contrast, the pull of the Moon, which opposes the gravitational force of the Earth persists in an amplified form. In all of its revolutions around the Earth, the research object is still subject to this force, at varying intensities.

The location with respect to the Moon alters constantly during every orbit of the research station around the Earth, resulting at relatively short intervals in a corresponding fluctuation in the gravitational forces operating in the space laboratory. The pulse of the force fluctuations thus no longer occurs every 6.2 hours. However, the experiments with the bean plants have demonstrated that the plants do not respond to every fluctuation, but are only in a position to do so once a movement already begun has been completed. The behavior of the mold fungus could be similar. The increased number of signals occurring in the space laboratory would, thus, not necessarily have any effect on the period of the rhythms, because the fungus would be ready to receive a signal leading to another reaction only after it has completed its spore formation. For the temporal configuration of a daily cycle, it is thus irrelevant that the impulses, most of which in any case are not utilized by the organism, occur at shorter intervals than on Earth.

2. In the natural environment, the duration of the rhythms corresponds exactly to the period of the regulatory rhythms in the environment. In the open air, the leaf movements of the bean plant follow the daily fluctuations of outdoor light exactly. The situation changes when the organism executes its rhythms under constant conditions in the research laboratory. The length of the period then varies, becoming shorter or longer, and no longer corresponds to the temporal cues given by the external factors. With respect to their duration, these "circa-rhythms" become autonomous and independent, developing their own dynamics.

Birds that breed in a 12-month annual cycle in their natural environment sit on their eggs every 9 to 10 months when placed in a cage under conditions of unvarying light and temperature (Farner and Lewis 1971). This dissociation of the length of the period from the external rhythms is also regarded as evidence for the existence of an internal chronometer, which, under constant conditions, "assumes responsibility" for determining the rhythm.

However, the experiments with the bean plants have shown that the rhythms generated by means of fluctuations of the tidal force appear "as an alternative" only when the light/dark cycle of daylight is artificially replaced by continuous illumination at a constant intensity. The fact that the rhythms then generated by the "lunar clock" have a different period than the leaf movement cycles controlled by fluctuations of sunlight is to be attributed solely to the differing (astronomically determined) temporal cues of the pacemakers, the Sun and the Moon; this deviation is therefore due to the nature of the factors involved. The discrepancy between the length of the periods displayed in the laboratory and in the open air thus poses no obstacle to the search for further "lunar clocks."

3. Chronobiologists also state that the rhythms generated by the "internal clock" are genetically determined and thus innate, with the length of their periods being inherited in accordance with Mendelian laws. In support of this, experiments are cited in which the eggs of stonechats are brought into the laboratory, and the young birds are raised from the egg under constant conditions. Because these animals also exhibit a breeding rhythm with an approximate annual cycle of almost 10 months, it is concluded that this behavior must be innate (Gwinner 1996).

However, it is possible that a "laboratory rhythm" with a period of approximately 10 months could also be an externally controlled "lunar cycle" several months long. In this case, the innate properties of the organism would consist only of the ability to react to the fluctuations of the tidal force, processing these signals and translating them into the relevant behavior. The influence of a "lunar clock" would likewise begin at birth.

In addition, scientists have cross-bred bean plants under constant experimental conditions in order to determine whether the varying periods of the circadian rhythms are transmitted to the next generation of plants according to the laws of genetics. Bean plants with a leaf movement rhythm averaging 23 hours under constant lighting conditions were crossed with plants exhibiting a cycle averaging 26 hours. In the next generation of plants, the length of the period of a complete leaf oscillation amounted to 25 hours. This value fell between the values for the plants in the parent generation, thereby conforming to one of the laws of genetics. It was concluded that the cause and the varying lengths of the circadian rhythms in the bean plants must, therefore, be genetically determined (Simons 1994).

The experiments with bean plants already described demonstrate unequivocally that in the absence of the light/dark cycle of daylight, the circadian rhythms often consist of two half-cycles differing in length; this depends on the intervals at which the bean plant reacts to the fluctuations of the tidal force. In principle, the possible time spans are 6.2, 12.4, and 18.6 hours. The average duration of a complete oscillation thus alters almost constantly from one series of experiments to the next, because it is dependent on the frequency with which the plant selects particular time spans for the temporal configuration of its oscillations. Thus, the emergence of different period lengths in breeding experiments can be entirely a matter of chance. In addition, the dependence of the circadian rhythms of the bean plants on the position of the moon has already been demonstrated experimentally.

4. Another circumstance supporting the argument in favor of the existence of self-generated diurnal rhythms is that for one and the same organism, the duration of these cycles under constant conditions can vary constantly in an irregular manner, spontaneously becoming longer or shorter.

Even organisms of the same species can display rhythms of completely different lengths in the laboratory. If the rhythm were determined by an unknown external environmental factor, it would be expected that these "individualistic" reactions would not occur; instead, all of the organisms involved would respond to the external signal "in step" with one another.

The answer to this argument has already been given. The "partial group behavior" of the bean plants during the experiment described demonstrates that each plant determines the period length and the duration of the oscillation of its leaf movements "at will." Although the signals, comprised of fluctuations of the tidal force, occur in a constant rhythm, each individual plant generally observes and responds to them only at irregular intervals. For this reason, the constantly varying period lengths and the consequent apparent mutual independence of the organisms are by no means compelling arguments for the existence of an "internal clock." The operation of a "lunar clock" could give rise to similar irregularities in the rhythmic processes.

5. The discovery of the clock genes in the cells of the mold fungus *Neurospora*, the fly *Drosophila*, and the mouse was a milestone in the field of chronobiology. In the view of the scientists, this discovery substantiated the genetic origin of the internal diurnal clock. The time was "fabricated" in the individual cells. The synthesis and breakdown of messenger material and protein in the cell takes place in rhythmic cycles, the duration of which corresponds approximately to the length of a day. This constituted a self-generating circadian rhythm in the smallest possible space.

However, as already mentioned, the problem of how the clock gene in its turn "measures" the time has yet to be solved. Who or what ensures that the whole process takes place at regular intervals? Because there is as yet no well-supported scientific theory to explain this, the clock gene is only referred to as at least a "preliminary" explanation of the basic mechanism of the "internal diurnal clock".

Nevertheless, it is at least conceivable that the fluctuation of the tidal forces may also exert an influence. These might possibly be able – as already hypothesized in the case of the bean plant – to influence the cellular environment and by this means to regulate the timing of the whole process. A redistribution of ions or other changes in the cell could just as easily be brought about in a rhythmic manner by this pacemaker. This periodically recurring external influence would then determine the beginning and end of the reaction.

6. The fly *Drosophila* was the experimental organism that provided additional insights into the "internal clock," considered by scientists to be genetically determined. If the clock gene of the fly was altered or lacking, breakdown or failure of the rhythmic configurations resulted.

Drosophila normally emerge from the pupa in approximately 24-hour cycles, and their mating calls are given at 56-second intervals. In some cases, genetic alterations led to the complete disappearance of recognizable rhythms. In other cases, they resulted in the period of the circadian rhythms of emergence of adult flies being lengthened to an average of 28 hours or shortened to an average of 16 hours. At the same time, the length of the pauses between mating calls was also altered.

Accordingly, there must also be a direct connection among the diurnal rhythm, the mating call cycle, and the clock gene.

The control center for the "internal clock" in human beings and other mammals is located above the intersection of the two optic nerves. In rats, experimental damage of this tissue caused the animals to lose all rhythmic reactions. For chronobiologists, such examples are a further indication that the "internal clock" is genetically based and controlled from within the organism.

With regard to both of these cases, it should be noted that the operation of the "lunar clock" discovered in the bean plant can be compared to the system of a remote-controlled radio clock. Any alterations or damage affecting the clock mechanism as a receiver of signals can, of course, also affect the operation of individual functions, even though temporal cues are still being emitted punctually from a remote transmitter.

Only if the clock stops completely or if there is a change in the times indicated by the clock can definite conclusions be drawn as to the nature of the control mechanism. It is thus possible that the interventions carried out in such experiments result merely in an impairment of the organism's ability to process the signals, so that although the signals continue to arrive regularly, they can no longer be put to use. Hence, these experimental results cannot serve to provide convincing evidence of a self-generated control of the rhythms.

From the foregoing, it can only be concluded that in light of the scientific discoveries that have so far been made, the existence of other "lunar clocks" cannot safely be ruled out. The "lunagent rhythms" of the bean plants are, thus, not necessarily to be regarded as a unique product of evolution.

It therefore seems justifiable to ask whether there are any indications or even evidence of a direct influence by tidal force fluctuations generated by the Moon on other biological rhythms that until now have likewise been attributed to an "internal clock."

As has already been established, in the natural environment the duration of the biological cycles of organisms corresponds exactly to the period of the environmental rhythms (e.g., light/dark or high tide/low tide) to which plants, animals, and human beings must adjust. In the view of chronobiologists, the "internal clock" itself at first generates "imprecise" rhythms, which are temporally altered and adapted to the fluctuation of environmental influences only by means of the decisive external conditions. In the absence of this adaptation, the inexact "circa-rhythms" of the "internal clock" again appear.

If the "inaccuracies" of the various types of "circa-rhythm" appearing in the laboratory are compared with one another, a peculiarity comes to light. In the bean plant, the duration of a complete leaf oscillation under constant conditions averages approximately 28 hours. Thus, the deviation from the 24-hour light/dark cycle of daylight amounts to an average of 4 hours. Birds breeding in cages in the laboratory sit on their eggs at intervals of somewhat more than 9 months. In their natural environment, they would carry out this activity at 12-month intervals. In these two cases, as well as in many others, the rhythms brought about by the "internal clock" deviate considerably from the periods of the relevant environmental cycles.

For human beings, the values are not quite so extreme. In conditions of isolation, removed from the influence of external pacemakers, the diurnal rhythms lengthen to approximately 25 hours.

All of these examples have one thing in common: In the natural habitat, the diurnal and annual cycles of the organisms are adapted to environmental conditions exclusively by means of the fluctuations of incident sunlight.

The situation is quite different in the case of organisms that in their natural habitat are controlled by the high and low tides at the coasts. In the laboratory, in the absence of these time givers, the periods of the rhythms of the "internal clock" generally correspond, just as before, to the intervals between high and low tides and, in particular, correspond exactly to the spring and neap tides at the coast. In contrast to life-forms that are oriented to light in their natural environment, the duration of the cycles of marine organisms does not alter in the laboratory. Precisely because of the correspondence of the period of the laboratory cycles with the duration of the rhythms found in the natural habitat at the coast, the results obtained from these experiments initially seem to point to a continuation of the tidal rhythms. This at first gives the impression that there are precise and imprecise "internal clocks."

In order to be able to ascribe the characteristic feature of imprecision to the "internal clocks" of life-forms oriented to the changing tides, scientists sometimes calculate the time in minutes. Thus, as soon as the period of the rhythms of marine organisms in the laboratory deviates by as little as 5 minutes from the target value of 24.8 hours (the time period from one high or low tide to the next but one), the timing is said to be imprecise.

However, this fails to take into account the characteristics of the astronomical phenomena. The target value of 24.8 hours is, in fact, only an average value. In its orbit around the Earth, the Moon follows an elliptical path, with the result that it travels through space at a variable speed with respect to the fixed stars. This causes variations in the intervals between the tides at the coasts. Thus, for example, a divergence from the average value of as much as 15 minutes per day can occur. Even cycles with a period of 25 hours and a few minutes occurring in the laboratory can still fall within the range of the duration of the natural tidal rhythms at the coasts and is not to be attributed to the inaccuracy of an internal chronometer (Schäfer 1988).

Nevertheless, there is a peculiarity that is generally exhibited in the cycles of organisms living at the coasts after they are brought into the laboratory. Although the duration of the rhythms does not fundamentally alter, the periodically recurring behavior patterns no longer appear simultaneously with the occurrence of high and low tides at the coast. The natural and laboratory rhythms have periods of the same length, but their phases are displaced with respect to one another.

However why does the "internal clock" operate so imprecisely in the laboratory relative to the rhythm of the natural environment in the case of organisms oriented in their natural surroundings to the light/dark cycle of daylight? In contrast, in the case of marine organisms living at the coasts, why do almost no discrepancies appear between the period of the cycles in the laboratory and the changing tides at the seashore?

For this, there could be only one explanation encompassing all of the phenomena described earlier, namely that it is not an "internal clock," but the Moon alone that determines the course of the rhythms in the laboratory.

In the natural coastal region, an organism would be dependent on orienting itself to the changing water levels of the high and low tides in its environment, so as to make optimal use of them. If the marine organism is then removed from its original habitat and placed under constant experimental conditions, in the absence of the high and low tides the "lunar clock" would assume control, in accordance with the principles demonstrated in the experiments with the bean plants.

However, the fluctuations of the astronomical tidal forces brought about by the terrestrial satellite do not coincide exactly with the rise and fall of the water level at the seacoasts. Currents, the contours of the ocean floor, and projecting land masses hinder the flow of water to the coast and also delay the subsequent ebb. For this reason, high and low tides appear at the seashore some time after the fluctuations of the tidal force that gave rise to them have already passed. A time lag occurs between cause and effect.

A marine organism that is brought from the seashore into the laboratory would thus be forced to respond to the astronomical tidal forces of the Moon acting on it directly, with no time lag. The duration of the rhythms exhibited in the aquarium would not alter relative to the cycles of high and low tides at the coast, because in both cases the rhythms continue to be based on the changing position of the Moon. Nevertheless, the phase of the rhythm displayed in the laboratory would change with respect to the time of high and low tide's at the ocean shore, because the effects of the tidal forces are exhibited immediately by the test subject, but appear only after a delay in the region of the seashore.

Quite different behavior is exhibited by organisms that orient themselves in their natural habitat to changes in day length or to the rising and setting of the Sun. In the laboratory, they are deprived of these cues and must orient themselves according to the "lunar clock." This indicates units of time that no longer have any connection with the presence of sunlight outside the laboratory. A change in the period of the rhythms necessarily results, because the signals of the Sun and the Moon occur at different intervals.

There are a few marine organisms that not only maintain the rhythm of high and low tides under constant conditions at a biological institute but also reflect other features of the tidal rhythms. The "beach hide-and-seek" behavior of the sand flea that inhabits the Californian coast has already been described. At high tide, the sand flea leaves its stony burrow and becomes active in the water. At low tide, it disappears again underground. In the aquarium, it also appears and disappears, in a rhythm corresponding to the tidal cycles. As soon as a spring tide occurs at the coast, the sand flea in the laboratory increases its activity, without apparently being able to sense this event. Accordingly, the "internal clock" must not only continue to "tick" in the rhythm of the tides, but, in addition, it must be able to provide the sand flea with information concerning a higher than average tide occurring beyond its reach on its "native coast." At all events, in the scientific literature this peculiarity is described as "remarkable." The perceptivity of this animal is indeed explicable only by means of the existence of a "lunar clock," which allows signals to reach the sand flea even in the laboratory.

A small North American fish of the cyprinodont family spawns approximately every 14 days, always at the time of a spring or a neap tide. Several fish were placed in each of a number of different aquariums under conditions of constant temperature, light, and water level. For the duration of the 4-month experiment, eggs were laid simultaneously in almost all of the aquariums, at times which coincided with the tidal extremes – the spring and neap tides (Waterman 1990). The scientists hypothesized that this synchronized timing, extending to all of the fish being held under different conditions, could perhaps be attributed to a signal distributed among the fish by means of the common water supply for the aquariums. The theory of the "internal clock" could not account for the complete temporal coordination of the joint spawning behavior in all of the fish tanks; only the assumption of a "lunar clock" could prove helpful here.

The unicellular alga *Gonyaulax* has already been described (von der Heyde, Wilkens and Rensing 1992). Even under laboratory conditions, this organism displays three different rhythmic processes occurring at different times: the emission of light, photosynthesis, and cell division. All of these events recur at intervals of approximately 24 hours; they are thus circadian cycles. Are there three "internal clocks" in one cell, or is one clock able to generate three rhythms differing in phase?

These findings are enough to indicate that the theory of the existence of a chronometer within the organism, implying an internalization of the biological rhythms, cannot account for the broad spectrum of multifaceted occurrences that have been observed.

The above-cited examples can be reduced to a common "denominator" if it is assumed that the "lunar clock" exhibited by the bean plants has a more far-reaching role as a pacemaker.

Even in the laboratory, the sand flea receives a stronger external signal at the time of a spring tide, because the spring tides are based on a change in the effect of the tidal force. Similarly, the small fish in the aquariums detect the impulse together, and, consequently, all react simultaneously.

The "lunar clock" could also make possible the temporally shifted cycles of the alga. This has already been indicated in the experiments with the bean plants, because each individual plant can give preference to either an H or an L position of the

Moon for the execution of a change in the direction of its leaf movements. In the case of the alga, if the emission of light occurs, for example, at the time of a fluctuation of the tidal force corresponding to an H position of the Moon, the photosynthesis, in contrast, could occur in the rhythm of the L positions. Although both cycles are circadian, there would be a time lag between them; in other words, their phases would be displaced with respect to one another. The cycle of cell division could then make use of a third circadian "mixed rhythm" consisting of a combination of H and L positions.

At the beginning of the 1950s, American biologists at Northwestern University in Illinois investigated the temporal behavior of oysters. They obtained these mollusks from a harbor basin located 1500 km to the west and placed them in an aquarium. The animals were then kept in continuous darkness (Palmer 1995).

In order to study their feeding behavior, it was necessary to record the times of the opening and closing of their shells. However, at first an inexplicable deviation occurred. The mollusks adopted a new time schedule for the regular opening and shutting of their shells. When the records of 4 weeks of observations were analyzed in detail, the researchers found that, without a doubt the mollusks in the laboratory were oriented exactly to the rhythm of the tides which would have occurred on the university grounds if there had been an ocean there.

This experimental example is regarded as "puzzling," but is scarcely discussed any more in the scientific literature concerning the "internal clock." It would also be inconsistent with the views of modern chronobiology. In addition to a "time sense," the "internal clock" of the oyster would also need to possess a "place sense" in order to be able to inform the organism of its present location so that it could adapt itself "according to the tides." For this reason, the internationally recognized American biologist *BROWN*. repeatedly called into question the existence of an internal chronometer and advised scientists to continue the search for a still unknown, rhythmically occurring external factor.

Once again, the behavior of the oysters in Illinois would be explicable in terms of the concept of a "lunar clock." The fluctuations of the tidal force are dependent on the positions assumed by the Moon relative to a particular point on the surface of the Earth. Because the Earth rotates from west to east beneath the Moon, places located at different degrees of longitude "pass" the Moon at different times. For instance, if the Moon reaches its highest point and thus assumes an H position relative to Berlin, it will assume the same position relative to Bonn, located to the west, about half an hour later. This means that the "lunar clock" "strikes" at different times in places located at different degrees of longitude.

When the oysters were brought from the coast to a research location 1500 km to the east, they were thus subjected to a different "lunar time." Hence, they adapted themselves to the decisive lunar positions relative to the place where the university was located. At the same time, this behavior reveals that the mollusks were orienting themselves according to the actual position of the Moon. Consequently, the Moon must exert a direct influence on one biological rhythm of these animals. This, in turn, requires that the oysters have the ability to perceive varying intensities of the tidal forces in their external environment.

Biological functions most frequently occur in diurnal rhythms, with a circadian period said to be generated by an "internal clock." Support for the existence of such a self-regulating diurnal clock has been found in the cells of mold fungi, flies, and mice, where the clock gene sets recurring processes in motion.

However, rhythms having a period of approximately 1 week, 14 days, 1 month, or 1 year also appear under constant laboratory conditions. Such regularly recurring events having periods longer than a day are also attributed by chronobiologists to the existence of an "internal clock." However, any conception of where in the organism such a chronometer could be located or what mechanism might underlie the longer-term temporal cues is lacking.

The Sun-Earth-Moon system yields a whole range of tidal force cycles with a great variety of intensities and period lengths. The tidal forces arise from the combined effect of centrifugal forces on the Earth brought about by its rotation with the Moon about a common center of gravity, and the gravitational attraction of the Sun and the Moon.

Although the Sun has a significantly greater mass than the Moon, the gravitational pull of the Sun on the Earth is barely half as great as that of the Moon. This is due to the fact that the Sun is too far away for the effect of its greater mass to be felt.

When the two heavenly bodies "pull together," the attractive forces originating from them and affecting the Earth are thus reinforced and strengthened. Such a situation occurs whenever the Sun, the Earth, and the Moon are positioned in a straight line. The phase of the Moon reveals when this takes place. At the time of a full or a new moon, the "triumvirate" is aligned approximately in a straight line.

At the time of a half moon, the Sun and the Moon form almost a right angle with the Earth. The two heavenly bodies then influence the Earth from different directions, with the result that the combined effect of the gravitational attraction is reduced to almost one, third of its former value.

From full moon to new moon to full moon is about 29.5 days, which corresponds to approximately 1 month. For various reasons, this value can vary by a matter of hours. The time from full moon to new moon amounts to around half of this time span, or somewhat more than 14 days. Every 2 weeks, the tidal forces are therefore stronger than average, as can be seen from the spring tides at the coasts.

There is also a second system, independent of these factors, which brings about fluctuations in the effects of the tidal force. The Moon orbits the Earth following an ellipse like path; a single orbit takes somewhat more than 27.5 days. Thus, in the course of its orbit, the distance of the Moon from our planet varies. At the two furthest points, the distance averages 405,504 kilometers; in between, the Moon reaches its two nearest points, at a mean distance of 363,296 kilometers from the Earth.

From the time when the satellite is at its maximum distance from the Earth (apogee), to the time, barely 7 days later, when it reaches its closest position (perigee), the intensity of its tide-generating forces varies by about a third.

Because the two lunar rhythms described above do not coincide with one another, their combined influence varies. At intervals of several months, their combined effect reaches a maximum. At this point, the Sun, the Earth, and the Moon are in a straight line, and at the same time, the Moon is at its closest to the Earth. The maximum effects of the different systems are thus added together, resulting in an increased influence of the tidal forces. The opposite is true when the minimum effects of the two systems coincide.

The constantly changing angular distance of the Moon from the equator also influences the intensity of the tidal force. There are also additional lunar rhythms, which could not, however, be expected to play any significant role with respect to the tidal forces (Schäfer 1988).

The Sun, the Moon, and the Earth continuously alter their positions with respect to one another, with the result that the distances between them constantly increase or decrease. This complex interplay of events also gives rise to a multitude of rhythmically recurring fluctuations of the tidal force, varying greatly in magnitude and period length.

The "lunar clock" is thus able to indicate numerous shorter and longer units of time. In order to generate regular cycles having a particular period, an organism would have to orient its "receiver" accordingly, in order to select the "preferred rhythm program" from the "rich assortment" of recurring signals.

In addition, the strength of the tidal forces is perpetually changing. They slowly increase in intensity and then gradually diminish again. This makes available an uninterrupted supply of energy that can be put to use. The tidal power stations on the coasts provide an example of the utilization of this type of perpetual energy supply. Both the incoming and receding waters unceasingly drive the turbines, generating electricity.

As mentioned earlier, the "lunar clock" is also able to regulate rhythms that are asynchronous with one another. Cycles exhibiting such phase displacement have already been observed in the case of the bean plants.

The next step is, therefore, to examine the various units of time specified by the "lunar clock," in order to determine a possible correspondence with the period lengths of the individual lunar rhythms.

The remarkable Indian telegraph plant *Desmodium motorium* is native to India and the Philippines. At the end of its leaf stalk are three individual leaflets: a longer one in the middle, and a shorter one on each side. The middle leaflet rises and falls in a diurnal rhythm – like the leaves of the bean plant – following the light/dark cycle of daylight. In contrast, the two lateral leaflets execute continuous circling motions, with a single "gyration" taking about 5 minutes. Both processes are controlled by means of an articulation. The ceaseless rotation of the lateral leaflets can be followed with the naked eye. The plants continue these movement cycles even under conditions of continuous illumination in the laboratory (Hensel 1993).

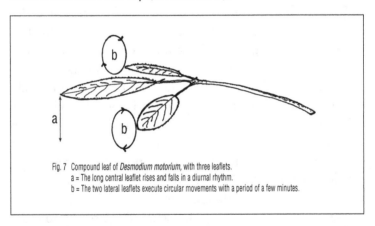

Fig. 7 Compound leaf of *Desmodium motorium*, with three leaflets.
a = The long central leaflet rises and falls in a diurnal rhythm.
b = The two lateral leaflets execute circular movements with a period of a few minutes.

Scientists have not yet been able to explain how the "internal clock" can regulate the diurnal rhythm of the central leaflet while making available the energy necessary for the continuous circular movements of the lateral leaflets.

As has already been discussed in connection with the leaf movements of the bean plants, the circadian rhythm exhibited by the central leaflet of the telegraph plant under conditions of continuous illumination represents a "lunar clock time unit." At the same time, the circling of the small leaflets at short intervals could be affected by an uninterrupted utilization of energy resulting from the continuous strengthening and weakening of the tidal forces, in such a way as to bring about the rotations. Accordingly, the central leaflet must make use of sensors differing from those used by the lateral leaflets, in order to respond only to the 6.2-hour rhythm of the alternations of the H/L positions of the Moon, resulting in a "circadian rhythm." The lateral leaflets, on the other hand, could have evolved a sensitivity to the continual changes in the tidal forces.

Chronobiologists also regard waking and sleeping in human beings as a 24-hour diurnal cycle adapted to external lighting conditions, which is generated by an "internal clock." Experiments have shown that when people are placed in isolation under constant environmental conditions, the "internal clock" causes the period of the rhythm to deviate to approximately 25 hours.

It should again be kept in mind that this "circadian" time span of 25 hours corresponds to the time required for a point on the surface of the Earth to revolve about the Earth's axis and arrive back at its initial position relative to the Moon. If the Moon were stationary with respect to the Earth, as the Sun is, the point would require only 24 hours to accomplish this, because it would have reached its goal after one full rotation of the Earth. However, because the Moon slowly travels in the same direction as the Earth's rotation, when a point on the globe that has rotated past the Moon then returns 24 hours later, the Moon is no longer in its original position. In the meantime, the terrestrial satellite has moved on, so that the point must cover a distance greater than one revolution of the Earth in order to overtake the Moon. For this reason, the lunar day is an average of 24.8 hours long, but can occasionally be as long as 25 hours and a few minutes.

The lunar day thus includes four tidal force fluctuations, which occur at intervals of 6.2 hours on the average. The "circadian" sleep/wake rhythm of human beings controlled by an "internal clock" therefore has a period corresponding to a lunar day; the light/dark cycle of the solar day, in turn, brings about the adaptation to the 24-hour rhythm of the environmental conditions.

In this connection, it is appropriate to refer back to the report of American doctors on the sleep behavior of a blind man. This patient was unable to adjust his "internal clock" to the 24-hour day, because he could not perceive the fluctuations of sunlight (Moog et al. 1990).

As has been shown by various experiments, blind people are also able to synchronize their internal chronometers to the length of a day by making use of other external factors to orient themselves to the particular time of day. Such factors could include sounds or the behavior of fellow human beings. In the case of this patient, such signals from the environment seemed to miss their mark, so that in his daily life, the period of his biological rhythms was determined only by his "internal clock." In the absence of correction by external factors, the sleep/wake behavior had a period of 24.8 to 24.9 hours.

Several other bodily functions, such as temperature, alertness, and cortisone distribution were also found to be based on these time periods. This meant that the patient's body had lost its temporal connection with the solar day; the blind man was sometimes tired in the daytime and alert at night. In addition, in the sleep laboratory, the doctors treating him noticed that throughout the investigation, over the course of several months, the beginning of sleep always coincided with the occurrence of a low tide in the nearby ocean.

The "internal clock" of the patient indicated the rhythm of a lunar day. The regular habit of falling asleep at low tide could be coincidental. However, in light of the long duration of the investigation, this seems unlikely. Rather, the synchronization of the beginning of sleep with the occurrence of a low tide points to the presence of a "lunagent rhythm." The body of the blind man reacted to the L position of the Moon in a recurring cycle of approximately 24.8 to 25 hours.

In studying the treatment of sleeping disorders, scientists also found that the "internal clock" could be reset by exposing the test subject to bright light for 4 hours. However, it was not enough to use any amount of light, at an arbitrarily selected time. It was found that, for inexplicable reasons, the duration of the exposure to light must be neither too short nor too long, and the exposure must occur only at certain unpredictable times. If this was done, the sleep/wake rhythm and the cycles of bodily functions were regularly shifted by about 6 hours. The light treatment thus caused the phase of the biological rhythms to be displaced by a time span corresponding to the interval between two fluctuations of the tidal force. In contrast, the wrong amount of light administered at the wrong time had no effect. American researchers are particularly concerned with clarifying the details and underlying causes of this phenomenon.

Surprisingly, this method can also be used to reset the "lunar clock" of the bean plant by 6.2 hours or a multiple thereof. If plants that have previously been grown under constant conditions are exposed to several hours of intense illumination by plant lighting, this can result in their leaves being raised toward the light. After termination of the illumination, the plants again continue to execute their lunagent leaf movement cycles. The light treatment temporarily suspends the operation of the "lunar clock." Once the interruption is over, the "lunar clock" again assumes control, operating in the rhythm of the fluctuations of the tidal force.

The phase of the newly recommenced lunagent rhythms is generally displaced relative to that of the rhythms that occurred before the intense illumination was administered. The phase displacement always amounts to one or more times the 6.2-hour interval between two fluctuations of the tidal force. The timing and duration of the exposure to light are also critical. The bean plants are not sensitive to light at all times. They generally react only when an individual leaf movement has been completed – in other words, when the leaves have already been in a horizontal or vertical position for some time. During the raising and lowering of the leaves, and also shortly after the completion of a movement, the cells that receive the signals are in a refractory period during which further stimuli, such as a subsequent tidal force fluctuation or an additional dose of light, have no effect.

In addition, in the few relevant investigations that have been carried out, it appears that a tidal force fluctuation must be pending in order for the plants to be sensitive to the disruptive illumination. Thus, the experiments show that the "lunar clock" of the bean plants is not always susceptible to influence and can be reset only by an average of 6.2 hours or a multiple of this factor.

Scientists have exposed human research subjects to interior lighting six times brighter than normal (Lewy 1983). Measurement of body temperature then indicated fluctuations with a period of 18.5 or 31 hours, a period length already familiar from the leaf movement rhythms of bean plants. Once again, the rhythm of the "lunar clock" is apparent, with periods that are multiples of the factor 6.2.

In addition, bean plants exhibit a behavior that could provide a possible explanation for what is referred to as "winter depression." If the plants are grown in the winter months and placed on a windowsill in dim daylight, under a sky completely covered with clouds, the leaf movements no longer follow the light/dark cycle of the insufficient outdoor light. Instead, the plants orient themselves – as in the laboratory – to the "lunar clock."

The reversal movements are executed basically in synchrony with the fluctuations of the tidal force occurring at the particular location. This can have the result that in the mornings, in weak light, the leaves remain vertical, whereas in the dark evenings, hours after the Sun has set, the leaves are still stretched out toward it.

This demonstrates that even in daylight, if the intensity of the light is insufficient, the "lunar clock starts up," with the result that the rhythms of the plants can no longer be optimally adapted to the actual conditions of the external light/dark cycle.

Effects similar to those associated with night-shift work can also be observed in bean plants. After the plants have executed leaf oscillations that follow the fluctuations of natural daylight, they are placed in artificial illumination at night and in a darkened room during the day. It takes at least a few days for the plants to adapt their movement cycles to the changed lighting conditions.

The "jet lag" experienced by travelers occurs particularly on flights from east to west, but also on flights from west to east. In the case of many travelers, the first symptoms often appear immediately after landing. These symptoms are attributed to the difficulty experienced by the body in adapting to the altered lighting conditions associated with the time change. Nevertheless, it is more frequently the case that people undergo such readjustments of their daily routine without immediately experiencing any noticeably harmful effects. The "lunar clock" might also be a cause of the disruption of physical and mental functions after arrival at the travel destination. When one travels to another time zone, not only the light/dark cycle of the solar day but also the timing of the H and L lunar rhythms are displaced. The corresponding signals from the terrestrial satellite are received at different times at different degrees of longitude. Therefore, under certain circumstances, it could be the "solar and lunar time changes" together that result in the undesirable effects that interfere with the pleasure of traveling.

A more recent discovery in the field of chronobiology is the 7-day rhythm of the human body. It is less strongly marked than the diurnal reactions and is expressed in slight changes in heartbeat, blood pressure, and resistance to infection (Halberg et al. 1986). At weekly intervals, the body is also particularly resistant to organ transplants. Injuries heal in 7 days. After one donates blood, the supply of red blood corpuscles regenerates in a weekly rhythm.

The 7-day cycles have one thing in common: There is no obvious environmental factor in the sense defined by chronobiologists (e.g., light fluctuations, temperature oscillations, etc.) that could correct an imprecise "internal clock" so that it maintains an exact weekly rhythm. It follows that there must be an exceptionally punctual internal chronometer that by itself can generate these longer-term weekly cycles: another "new clock"?

The garden bean also exhibits a 7-day rhythm (Sundararajan 1986). The water required for germination is not absorbed by the seeds at a constant rate. The absorption of moisture increases at weekly intervals, at the time of the full moon, half moon, and new moon. This has been substantiated by scientific experiments.

The week is also a time unit of the "lunar clock." There is always an interval of approximately 6.9 days between the time that the terrestrial satellite is furthest from the Earth and the time that it is nearest the Earth. These variations result in a greater than average decrease or increase in the effect of the tidal forces. Approximately every 7.4 days, the Sun also becomes involved in the interplay of forces. At these intervals, the Moon "dances" out of the straight line it forms with the Sun and the Earth and then back again. At approximately weekly intervals, the gravitational attraction of the Sun therefore likewise reinforces or reduces the effect of the lunar forces.

The 14-day cycle of marine organisms that follows the extreme tidal force fluctuations that generate the spring and neap tides has already been described.

As has been confirmed repeatedly by scientific investigations, in the human body, fluctuations in the excretion of uric acid coincide with the time of the full and new moon. At intervals of approximately 2 weeks, the uric acid level is particularly low, when the Sun, the Moon, and the Earth are positioned roughly in a straight line, as indicated by the phase of the Moon. Because the uric acid level is determined by several different metabolic processes in the body, it can be concluded that these processes are also based on this lunar rhythm.

This 14-day rhythm in human beings clearly follows the lunar cycle. The fluctuations of the weak light of the Moon must be excluded from consideration as a possible external time giver that could correct and adjust an imprecise "internal clock." In addition, experiments have shown that these 14-day rhythms occur even when the test subject does not set eyes on the terrestrial satellite illuminated by the Sun. Thus, in this case as well, no environmental factor recognized by chronobiologists can be identified that synchronizes the rhythms generated by the body itself with the changing positions of the Sun and the Moon.

If one accepts the basic scientific position that a direct influence of the Moon on human beings must be ruled out, one is forced to fall back on a hypothetical model of an internal chronometer with prophetic properties. In the course of evolution, such an "internal clock" would first have to store the exact time of the full and new moon. The times would have to be accurate to the nearest minute, because if the clock "miscalculated" each event by only as much as 60 seconds, in 60 years the internal chronometer would indicate the time of the full and new moon 1 day too soon or too late. Such precision can scarcely be assumed.

However, a decisive consideration is the fact that the speed of rotation of our planet, the orbital period of the Moon, and the distance of the Moon from the Earth have constantly changed over time. The time interval between the phases of the Moon is therefore not invariable. All of this would have to have been taken into account by the "internal clock" in order for the fluctuations in the uric acid level to be still synchronized exactly with the actual positions of the Moon in the sky. It is inconceivable that the human organism could calculate in advance this constantly varying arrangement of the three heavenly bodies: the Earth, the Sun, and the Moon.

Rather, it seems much more probable that the current positions of the triumvirate and the associated fluctuations in the intensity of the tidal forces are indicated to the human body from without, by the "lunar clock," and that the human body possesses the ability to receive the corresponding impulses, in order to make use of them for the temporal configuration of particular rhythms.

The breeding behavior of the African stonechat has already been discussed. In a cage, under constant environmental conditions, the birds breed at intervals of somewhat more than 9 months. In the natural habitat of these animals, the increase and decrease in day length over the course of the year prolongs the time span specified by the "internal clock" to a 12-month breeding cycle. This is the scientific explanation. In order to keep an exact record of the timing of the reproductive rhythm in the laboratory, among other things, the diameter of the testicle of a stonechat was measured at intervals over a period of more than 10 years. During the time of the experiment, the duration of the reproductive cycle averaged 9.3 months (Gwinner 1996). This corresponds to 10 lunar months, where the length of such a month can be variously defined from the astronomical point of view. On the one hand, based on the length of time from full moon to full moon (the so-called synodical month), 10 lunar months would come to approximately 295 days. On the other hand, based on the interval from the time that the Moon is nearest the Earth to the next time this occurs (the so-called anomalistic month), 10 lunar months would amount to around 276 days. The temporal relationship between the reproductive and lunar cycles was also recognized by the scientists. However, they apparently considered the light of the Moon to be the only possible lunar influence that could affect the birds. The researchers thus believed that they could eliminate any external influence of the Moon by keeping the stonechats in cages under conditions of constant illumination.

However, the tidal forces recognize no boundaries. Although there is no known unique lunar rhythm that generates tidal force fluctuations with a period of approximately 276 to 295 days, even if there is no such rhythm, such a cycle could consist of a combination of individual shorter-term lunar rhythms. It is conceivable that the birds could develop their reproductive readiness slowly, in several stages, with the duration of each of the individual stages being based on a lunar rhythm of tidal force fluctuations. This idea is supported by the fact that the period length of the complete cycle corresponds to a multiple of the known short-term lunar rhythms.

Scientists find the behavior of the grunion, which lives off the coast of California, both surprising and inexplicable (Waterman 1990). As already mentioned, these fish always spawn in the coastal sand three to four nights after the new or full moon, in the first 2 hours following high tide. This occurs at the time of the spring tides, which are brought about by the position of the Moon. This extremely punctual behavior cannot be attributed to an "internal clock."

Such an "internal clock" would have to have the ability to predict to the nearest hour the times of the relevant phases of the Moon, with all of their temporal fluctuations. Nor can the light of the full moon be considered to be the signal giver. On the one hand, the fish also react at the time of the new moon, when the sunlight striking the terrestrial satellite is not reflected to the Earth. On the other hand, the grunion could not be expected to have the ability to "read" the slowly altering phases of the Moon with a precision that is accurate to the day. It is conceivable that the greater than average changes in water level occurring at the time of a spring tide could trigger the behavior. However, how are the fish to sound the depth of the water? Intensified wave action, which could mechanically trigger a response, is not necessarily associated with a spring tide. The reproductive behavior of the grunion is, in fact, only comprehensible if one assumes that they have the ability to perceive the increased tidal force fluctuations directly, by means of the "lunar clock," in the manner demonstrated in the experiments with the bean plants.

The scientific literature contains a multitude of graphs that indicate the course of the rhythms brought about by the "internal clock" of a great variety of organisms under constant experimental conditions. On the one hand, the progression of the oscillations before and after an interruption by an external influence (e.g., an added dose of light) is frequently illustrated. After the interruption, the organism can be seen to resume its previously executed cycles in their original form; however, the timing of the cycles is generally shifted. On the other hand, the graphs often depict the rhythmic behavior of several members of the same species occurring at a given time. In this case, the duration and the progression of the individual cycles are essentially similar; however, the rhythms are executed at different times.

Sometimes, organisms at first react in synchrony with one another, but then later diverge, in groups or individually, along different rhythmic paths.

The graph diagrams of particular interest here are those in which the period length of the cycles does not alter, but the oscillations are displaced with respect to one another. These repeatedly observed phase displacements serve to provide scientists with further support for the argument that all of the "internal clocks" function independently of one another and are not subject to any type of uniform control from the environment. The magnitude of the phase displacement is rarely explicitly indicated. When it is, values of 90° or 180° are almost always specified. This means that the displacement of one rhythm with respect to another is equal to one-quarter or one-half of the period of the rhythm.

However, in most cases, one is merely instructed to examine the graphs. Here, a comparison again reveals that the phase displacement that occurs is apparently regularly one-quarter or one half of the period length. For example, the maximum of one curve coincides with the minimum of another, or an extreme value of one curve falls exactly midway between the two extreme values of a corresponding curve.

This, in turn, argues against the existence of an "internal clock." Such an individual chronometer would specify the rhythms for each organism "at will," with no regard for the rhythms of other clocks or for its own rhythm before an interruption. According to the laws of probability, this would mean that phase displacements of every magnitude would occur with equal frequency, with no preference being given to certain values. A fixed relationship between the cycles of organisms of the same species cannot be reconciled with the principle of individual time measurements being carried out by each organism.

The leaf oscillations of the bean plants brought about by the "lunar clock" have shown that not every incoming signal is used to regulate the rhythms. The individual plants do not always react together to a particular external impulse, but determine independently from one another which of the available signals they will use to begin and end their cycles. If different signals are selected, a phase displacement of the rhythms results. Because the signals of the "lunar clock" occur only at fixed intervals, each phase displacement must accordingly be based on prespecified values. In contrast, an independent "internal clock" would follow its own course, and in determining its phase, it would not orient itself to the rhythms of the clocks of other organisms of the same species.

There is another general point that should be considered. Insofar as the "internal clock" specifies circadian rhythms with a period corresponding only approximately to the length of a day, the period lengths for various living beings reported in the literature are found to average 18 to 30 hours. Thus, the internal diurnal clocks of the individual organisms not only deviate sharply from one another, but they are also very imprecise with respect to the target value of 24 hours.

The situation is different with respect to rhythms with a period length of 1 week, 14 days, or 1 month. These periods are maintained very precisely, and in synchrony with one another, by the "clocks" of all of the organisms involved. However, if the "internal clocks" of all of the various organisms cannot control and synchronize exactly rhythms with a period of a day, in the case of longer-term rhythms they would be expected to display a complete lack of synchrony and even more extreme fluctuations of precision.

This contradiction can, in fact, only be convincingly explained insofar as it is assumed that the rhythms are configured by the "lunar clock." Approximately every 6.2 hours, by means of the fluctuations of the tidal force, the terrestrial satellite provides a signal that the organism can use for the regulation of its cycles. The length of a day could thus be approximated by four tidal force fluctuations, amounting to 24.8 to 25 hours. However, as demonstrated by the experiments with the bean plants, these signals are used in only an irregular manner. This means that, in the establishment of a diurnal rhythm, if the "lunar day clock" does not "hit the mark" of the 24.8-hour to 25-hour rhythm, it will necessarily run at least 6.2 hours too slow or too fast, because the signals are given only at these intervals.

However, the "lunar clock" cannot give rise to this type of deviation in the period lengths of longer-term cycles. In the case of diurnal rhythms, several signals of the same type are available that can be used to arrive at a period of "circa" 24 hours. These signals arise from one and the same lunar cycle. For the generation of events that recur at intervals of 1 week, 14 days, or 1 month, organisms can react only to a single specific tidal force impulse that occurs exclusively at the relevant time. If this signal is not observed, there is no other signal of the same "type" in the temporal vicinity that could be followed in order to maintain at least approximately a period length of 1 week, 14 days, or 1 month. The next impulse is given only when the full time span has elapsed. Thus, the organism must respond according to the rule Now or Never.

Therefore, in contrast to the situation with respect to diurnal cycles, it is not possible for the "lunar clock" to give rise to longer-term cycles that have variable periods of only approximately the same length. Consequently, the cycles of 1 week, 14 days, or 1 month have the same fixed form for all living beings; whereas each individual could have recourse to different signals for the generation of diurnal rhythms.

Experiments have shown that the circadian rhythms of human beings are fairly constant, with a period of approximately 25 hours. This corresponds to the time required for four tidal force fluctuations, or one lunar day. Human beings thus seem able to react to the "lunar clock" more consistently than is the case with bean plants, which exhibit diurnal rhythms that are not always based on the same number of tidal force fluctuations.

In light of all of the above, is the theory of a self-regulating chronometer internal to the organism, which gives rhythmic impulses to plants, animals, and human beings at intervals of days, weeks, and months, still tenable?

The data so far obtained in the field of chronobiology do not correspond perfectly with the hypothetical model of an "internal clock." There are too many exceptions, discrepancies, and, above all, contradictions for the multitude of experimental results to be explained in a unified, comprehensive manner by means of this scientific theory. A few findings directly exclude the possibility of the operation of an "internal clock"; for example, the oyster, which seems able to orient itself geographically, and the sand flea in the aquarium, which senses the spring tides of its remote native coast. This also applies to human beings, whose uric acid level is dependent on the current phase of the Moon.

A theory should be discarded if an entirely new, much more convincing explanation of the already known facts is available. The history of the natural sciences has shown that complex phenomena that express themselves in a great variety of ways are often explicable in terms of a few general principles.

It is, therefore, to be concluded that living organisms do not possess a self-regulating "internal clock" that specifies rhythms having a period length of approximately 1 day, or weeks or months. Such cycles appearing under apparently constant external conditions are brought about rather by the astronomical tidal forces, the intensity of which fluctuates at various intervals.

On the one hand, this conclusion is based on the experimental evidence that the circadian leaf oscillations of bean plants occurring under conditions of continuous illumination are based on the fluctuation of the tidal forces. It was these leaf movements that were decisive with regard to the scientific acceptance of the existence of an "internal clock."

On the other hand, a universal "lunar clock hypothesis" allows the individual complex processes and apparently contradictory facts to be explained in a convincing manner on the basis of a simple fundamental principle, without requiring special auxiliary hypotheses to explain exceptional cases.

The existence of "lunagent rhythms" also correlates well with our knowledge of the evolution of life, which began in the primeval oceans. External factors such as light and temperature were subject to irregular fluctuations at different depths in the ocean. They were, thus, not suitable for the creation of rhythms. In contrast to all of the other environmental influences, only the fluctuation of the tidal forces was present everywhere and exhibited its effects at constant intervals. Living beings that oriented themselves to this phenomenon thus made use of the only "precise clock," which gave them a considerable advantage.

Only when the blue-green algae began to employ solar radiation in the process of photosynthesis did light become vital for the further evolution of plants. The "tidal clock" was no longer sufficient and had to take second place to the "light clock," without, however, becoming superfluous and thus being "discarded" in the course of evolution.

Chronobiologists proceed on the assumption that the inexact cycles of the "internal clock" are synchronized and brought into temporal agreement by the rhythms of the environment. However, with regard to the "lunar clock," it seems more reasonable to assume that the environmentally controlled cycles and the "lunagent rhythms" represent two autonomous but not necessarily completely independent systems. The environmental signals of the light/dark cycle of sunlight and the high and low tides at the coasts have priority, because an adaptation to these factors is necessary for the survival of the organism. In the absence of these signals, or when there is no requirement for an adaptation to external conditions, the "lunar clock" comes to the fore.

This might also explain why the "lunagent rhythms" have been retained. If an organism is required to adapt itself to fluctuating environmental factors, such as the light/dark cycle of day light or the changing tides, it orients itself temporally to these events. In the laboratory, these controlling conditions are absent, a situation that could also arise under certain circumstances in the natural environment. Just as an emergency power generator begins to function in the event of a power failure, under these conditions the "lunar clock" then begins to operate, in order to prevent an interruption of "rhythmic activity." In such a situation, it does not matter whether the lunar cycles are coordinated precisely with the timing of the changing environmental factors, because these external influences are no longer present and there is thus no longer any necessity for following them exactly. All that matters is for the rhythms to be continued in a similar form, so that functions vital to the organism are maintained.

Under natural conditions, in the event of a temporary cessation or disruption of the recurring environmental phenomena, the "lunar clock" could thus serve as an "auxiliary pacemaker," rhythmically bridging the gap to enable the organism to survive this life-threatening situation.

On the other hand, the "lunar clock" can also still serve as the sole rhythm giver if no adaptation to other external factors is required for purposes of survival. This can be seen, for example, in the weekly rhythms not controlled by any other external influence and in the fluctuating uric acid level of human beings, which is dependent on the phases of the Moon.

It is obvious that the thesis of a "lunar clock" contradicts a fundamental premise of the field of chronobiology. Hence, the fate of this theory and the accompanying experiments is unfortunately already clear. Human beings, including scientists, tend to be involuntarily trapped by the past. We find it extremely difficult to call into question explanations that have once been believed and accepted on the basis of conviction. At least initially, the accompanying intellectual uncertainty and the possible necessity of admitting one's own mistakes prevent the experts from analyzing the new arguments; they are at first completely ignored. Under certain circumstances, this initial silence can be followed by a period of controversy, when the old theory is defended. In all probability, generations of bean plants – apparently guided by an "internal clock" – will raise and lower their leaves before the new theory is accepted.

Appendix I: Experimental procedure

In principle, all types of bean plant are suitable for the experiment. More precise results can be obtained under conditions of continuous illumination than under conditions of uninterrupted darkness. Nevertheless, in either case, the leaf oscillations are not carried out indefinitely. After a few days, the leaves cease to move, and the oscillations come to a standstill.

However, the individual plant varieties differ with regard to the length of time their leaf movements continue under constant lighting conditions. The runner bean *(Phaseolus multiflorus)*, especially the scarlet runner, is in this respect best suited for the experiment. Its movements often continue to be executed for a week or longer. However, this type of bean has the disadvantage that in the winter months, in insufficient light, it often scarcely follows the light/dark cycle of outdoor light, and its leaves move only a slight amount. This can be remedied by placing the plants under artificial plant lighting, which can be controlled by a timer. Light is to be administered beginning at 9 a.m., followed by darkness at 4 p.m.

The jack bean *(Canavalia ensiformis)*, varieties of which are sold under names such as "Pfälzer Juni" and "Hilda", is also suitable.

The bush bean "Saxa" *(Phaseolus vulgaris)* can likewise be used, particularly because when placed on a windowsill, it regularly executes its leaf movements even in dim outdoor light during the winter months.

The plants are to be grown in pots, in garden soil. It is advisable to remove the growing shoot between the first two leaves and to support larger plants by attaching them to a short stick.

As soon as the leaves are almost full grown and the plants have begun to raise and lower their leaves in synchrony with the light/dark cycle of daylight, the bean plants are ready for the experiment.

The plants are then placed under continuous illumination in a room that has otherwise been completely darkened. Because of the composition of their light, incandescent lamps are not suitable for plant lighting. White neon light ("white" or "cool white") is to be used exclusively as a light source. Movable fluorescent elements (18 watt or at most 36 watt) can be obtained in stores that sell aquariums. These lights are to be attached to a stand or other support in such a way that their height can be adjusted, in order to control the intensity of the light reaching the plants. The light intensity should be approximately 140 lux, measured at the articulation between the leaf and the leaf stalk.

If no photometer is available for measuring the light intensity, the following values can be used in order to adjust the height of the lamps:

Flourescent Lights	Watts	Distance from Plants
Osram universal white	18-watt/25	75 to 80 cm
Sylvania cool white	25-watt/30	80 to 90 cm
Sylvania white	58-watt/184	230 cm

In addition, a current astronomical almanac is required, which gives the time of the culmination of the Moon each day. This is the time at which the Moon reaches its highest point in the south, simultaneously assuming its upper H position.

However, the values given in the almanac are valid only for a particular location on the surface of the Earth (for instance,

10° east longitude and 50° north latitude). Because the Moon reaches its highest point at different times for different degrees of longitude, it might be necessary to use a conversion factor in order to obtain the culmination times corresponding the geographic location of the research site. Almanacs include location tables or conversion formulas for this purpose.

In addition, in an astronomical almanac, the times indicated almost always refer to the standard time of the particular time zone –, for example Central European Time (CET). Therefore, when statutory daylight saving time (or summer time) is in effect, 1 hour must be added to each of the values given in the almanac. In this way, the time of the culmination of the Moon is calculated for each day of the experiment.

In order to ascertain the times of the subsequent L and H positions of the Moon, the time between one culmination point and the next is divided by 4. An example might make this clear. For July 1, 1998, the time of the culmination given in the almanac is 6:20 p.m.; for the second day of the month, it is 7:02 p.m. The experiment is to be carried out in the vicinity of Heidelberg. According to the table in the almanac, for this location a time correction of +5 minutes is required. Because summertime is in effect, an additional hour must be added, with the result that the times given for July 1 and 2 are each put forward by 65 minutes. Thus, the decisive H position of the terrestrial satellite occurs at 7:25 p.m. on July 1 and at 8:07 p.m. on the following day. Between these two times is a time span of 24 hours and 42 minutes. A quarter of this span is thus 6 hours and 10.5 minutes. Therefore, the F position of the Moon that occurs at 7:25 p.m. on July 1 is followed 6 hours and 10.5 minutes later by the next L position of the satellite. Subsequently, at equal intervals, the moon reaches another H position, and then an L position, finally returning to its original upper F position with respect to the research site at 8:07 p.m. on July 2. Although the method of calculation used is not completely accurate from an astronomical point of view, it, nevertheless, meets the requirements of the experiment.

In order to measure the leaf movements and to determine the times of the turning points, at the appropriate times the distance between the leaf tips is recorded to the nearest millimeter. This is best done by means of a beam compass with a large span, but, if necessary, a simple ruler can be used. Alternatively, the distance of the leaf tips from the soil can be ascertained for both leaves.

Measurements are to be made approximately every hour, in such a way as to ensure that a measurement time coincides with the calculated time of occurrence of each of the tidal force fluctuations. Shortly before and also after these events, the measurements should be made at shorter intervals in order to obtain clear documentation of the turning points of the leaf movements, which usually occur at the exact time of a tidal force fluctuation, accurate to the nearest minute. For ease of analysis, the measurements recorded are to be transferred to graph paper, in the form of a curve.

In order to verify the exact correspondence of the turning points with the positions of the Moon, the bean plants are deprived of the fluctuations of daylight and subjected to uninterrupted illumination, using neon lights. It is advisable to carry out the investigation with two or three plants simultaneously. Measurements are to be made starting approximately 24 hours after the beginning of the experiment, because immediately after the change to continuous illumination, the bean plants sometimes have difficulty adjusting, often executing only short, irregular movements.

The decisive aspect of all of the experiments is the determination of the turning point, the time of the initiation of a reversal movement after a preceding upward or downward oscillation.

In the preceding Figures, for purposes of illustration, the leaf oscillations have been presented in a somewhat schematic form. In fact, the course of the movements is usually not as clear and uniform as it appears in these Figures. The plants raise or lower their leaves to the end position. They then often maintain this position for several hours, during which time, continuous slight, irregular oscillations occur, usually with the overall result that the leaves are lowered a little from the horizontal position or raised slightly from the vertical position. Only later is the decisive change of direction initiated.

The turning point is, thus, not to be confused with the moment at which the leaf tips of the bean plant happen to reach their widest or narrowest span. Hence, the extreme values of the curves do not necessarily coincide temporally with the turning points. A turning point is to be identified as the time of the initiation of a definite, uniform reversal movement, a movement that appears almost linear in form, without exhibiting any type of oscillation.

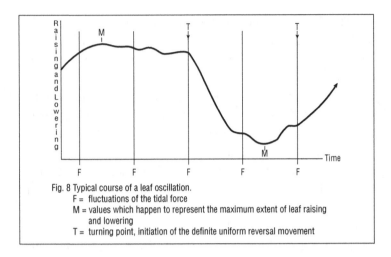

Fig. 8 Typical course of a leaf oscillation.
F = fluctuations of the tidal force
M = values which happen to represent the maximum extent of leaf raising and lowering
T = turning point, initiation of the definite uniform reversal movement

In addition, when plotting the curves, it is necessary to choose a sufficiently large time scale for the horizontal coordinate. A time span of 1 hour should be represented by an interval of at least 2 centimeters on the graph paper. Otherwise the diagram will appear compressed, and the subtleties and finer distinctions will not be sufficiently clear.

Turning points leading to descending leaf movements can generally be plotted more accurately than those leading to ascending movements. It is, therefore, advisable for the analysis to be restricted to the determination of the former, in order to verify their correspondence, accurate to the nearest minute, with the times of tidal force fluctuations.

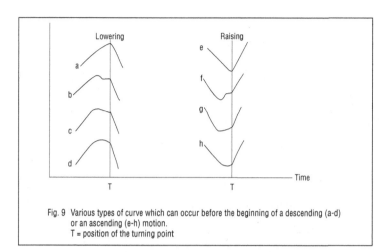

Fig. 9 Various types of curve which can occur before the beginning of a descending (a-d)
or an ascending (e-h) motion.
T = position of the turning point

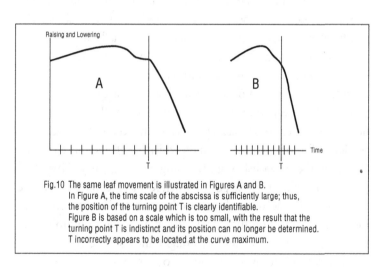

Fig.10 The same leaf movement is illustrated in Figures A and B.
In Figure A, the time scale of the abscissa is sufficiently large; thus,
the position of the turning point T is clearly identifiable.
Figure B is based on a scale which is too small, with the result that the
turning point T is indistinct and its position can no longer be determined.
T incorrectly appears to be located at the curve maximum.

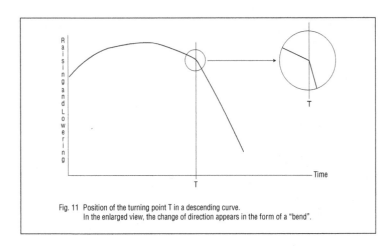

Fig. 11 Position of the turning point T in a descending curve.
In the enlarged view, the change of direction appears in the form of a "bend".

The reversal of a movement is frequently foreshadowed 45 to 60 minutes before the occurrence of a decisive lunar position. The plants seem to relax, in the sense that they gradually abandon their previously maintained day or night position and lower or raise their leaves a few millimeters in the contrary direction. In these cases, the exact turning point is not very easy to determine on superficial inspection. However, precisely at the moment of the fluctuation of the tidal force, the contrary oscillation is commenced in a considerably accelerated, more definite manner, which is reflected in a distinct "bend" in the previously slightly falling or rising curve. This is the precise time of the turning point, which regularly coincides with the lunar positions, accurate almost to the minute. The movement in the contrary direction then continues at a constant undiminished speed, represented by an almost straight line on the graph.

Plants do not always respond to signals with the same regularity as machines. Disruptions can occasionally occur. However, these rare deviations from the norm can be comprehensively described as follows:

1. Now and then, at first only one of the two bean leaves is raised or lowered at the time of a fluctuation of the tidal force. The other leaf is delayed, so to speak, but later suddenly follows, at a greater speed.

2. Both the leaves begin a slight reversal movement punctually at the time of an H or L position of the terrestrial satellite however, they suddenly cease to move. The leaves then remain in the same position for some time, seemingly held fast and not moving a millimeter. This is represented by an absolutely straight horizontal line on the graph. Then, the movement is suddenly continued, at a time that need no longer correspond with a fluctuation of the tidal force.

3. After the beginning of a rising or falling movement that is temporally correlated with a corresponding lunar position, the movement is abruptly broken off. The leaves move back again, and then later, at an arbitrary time, they complete the previously begun oscillation. In the case of this type of "retrogressive oscillation," the impression arises that the decisive "lunar force fluctuation" is at first insufficient to bring about a complete reversal movement, so that the leaves "spring back" again. Only later, after the fluctuation of the tidal force makes itself felt to an increased extent can the reversal movement that was broken off be completed.

However, even in the case of all of these observed disruptions relative to the course of a normal oscillation, it can be seen that the rhythm of the leaf movements of the bean plants is specified only by the astronomical fluctuations of the tidal force brought about by the Moon.

Appendix II: Supporting data

Regrettably, Dr. Gunter Klein died before he could oversee the final form of his manuscript. Also, at the time of his death, he was attempting to gather more data in connection with his hypothesis linking the positions of the Moon with the autonomic movements of bean leaves. Probably, Dr. Klein would have included some illustrative material of his work in this direction. After all, it is not feasible to propose a new theory without actually presenting any supporting data. Nevertheless, some indications of Dr. Klein's own researches on leaf movements are afforded by the sketches shown in Figures 8, 9, and 10 of Appendix I.

Fortunately, at least some of Dr Klein's results are still available for scrutiny and the original grpahs were kindly made available to the present author by his widow, Frau Klein, via his friend, Prof. Dr. Peter Becker, University of Marburg, Germany. The graphs show examples of leaf movements obtained from *Phaseolus vulgaris*, grown and observed under the conditions defined in Appendix I. Upon these graphs of the autonomic movements have been inscribed the F position of the moon. In many cases, there is a clear relationship between the time of attainment of the F position and the reversal of a leaf movement, just as shown in Figure 8 of Appendix 1. One representative graph of the leaf movement, together with the time of the lunar F position is shown in Figure 12.

In order to provide a further check on these data, lunar tidal forces were kindly computed by Prof. Dr. E. Klingelé (Swiss Federal Institute of Technology, Zürich) for the dates and location (Usingen, Germany) of Dr. Klein's observations. The graph corresponding to the data shown in Figure 12 is presented in Figure 13. The turning points in the lunar tidal force are recorded as arrowheads on the graph in Figure 12.

References

ARMSTRONG, S.M.: Melatonin and circadian control in mammals. In: Experientia 45. Cambrigde 1989, 932-938.

ASCHOFF, J./ DAAN, S./ GROOS, G.A. (Hrsg.): Vertebrate circadian systems. Berlin 1982.

BARGIELLO, T.A./ YOUNG, M.W.: Molecular genetics of a biological clock in Drosophila. In: Proceedings of the National Academy of Sciences of the USA 81. 1984, 2142-2146.

BERNDT, A./ GRAMATTE, T./ OERTEL, R./ TERHAAG, B./ RICHTER, K./ KIRCH, W.: Day-night variations in the renal excretion of the antiarrhythmic agent tiracizine and its metabolites. In: Chronobiology International 12. 1995, 135-140.

BOGENRIEDER, A. (Hrsg.): Lexikon der Biologie, Band 2, Stichwort Chronobiologie. Freiburg 1987.

BOGENRIEDER, A. (Hrsg.): Lexikon der Biologie, Ergänzungsband, Stichwort Chronopharmakologie. Freiburg 1987.

BROWN, F.A.: Response to pervasive geophysical factors and the biological clock problem. In: Cold Spring Harbor Symposia on Quantitative Biology 25. New York 1960, 57-71.

BÜNNING, E.: Die physiologische Uhr. Berlin 1977.

CHADWICK, D.J./ ACKRILL, K. (Hrsg.): Circadian clocks and their adjustment. Chichester 1995.

CHAN, H.A./ FOLK, G.E./ HUSTON, P.E.: Age comparison of human day-night physiological differences. Aerosp. Med. 39. 1968, 608-610.

DARWIN, C.: The power of movement in plants, New York, 1896.

DEANFIELD, J./ SELLIER, P.: Diurnal distribution in cardiovascular disease. Prognosis and therapy. Basel 1994.

DE CANDOLLE, A. P. : Physiologie végétale. In : Librairie Faculté de Médécine. Paris 1832.

DE MAIRAN, J. : Observation botanique. In : Histoire de l'academie Royale des Sciences. Paris 1729.

DOWSE, H.P./ HALL, J. C./ RINGO, J. M.: Cicadian and ultradian rhythms in period mutants of Drosophila melanogaster. In: Behavior Genetics 17. 1987. 19-35.

EDMUNDS, L. E. Cellular and molecular bases of biological clocks. Berlin 1988.

EHLERS, C. L./ FRANK, E./ KUPFER, D. J: Social zeitgebers and biological rhythms. Archives of general Psychiatry 45. 1988, 948-952.

ENDRES, K./ SCHAD, W.: Biologie des Mondes. Stuttgart 1997.

ENGELMANN, W./ KLEMKE, W.: Biorhythmen. Heidelberg 1983.

ERTEL, S.: Space, weather and revolutions. In: Chizevsky's heliobiological claim scrutinized. Studia Psychologica 38. 1996, 3-22.

FARNER, D. S./ LEWIS, R. A.: Photoperiodism and reproductive cycles in birds. In: Photophysiology 6. 1971, 325-370.

FELDMAN, J. F./ HOYLE, M.: Isolation of circadian clock mutants of Neurospora crassa. In: Genetics 75. 1973, 605-613.

GUTENBRUNNER, C./ HILDEBRANDT, G./ MOOG, R.: Chronobiology and chronomedicine. In: Basic research and applications. Frankfurt am Main 1991.

GUTZWILLER, M.C.: Der Mond in der menschlichen Geschichte, Sterne und Weltraum. Heft 8-9. 1996.

GWINNER, E.: Der innere Kalender tropischer Vögel. In: Biologie in unserer Zeit, Heft 3. 1996.

HAMNER, K. C.: The biological clock at the South Pole. In: Nature Nr. 4840. 1962.

HALBERG, E./ HALBERG, F/ HALBERG, J./ HALBERG, F.: Circaseptan (about 7-day) and circasemiseptan (about 3,5-day) rhythms and contributions by Ladislav Dérer. In: Biologia (Bratislava) 41. 1986, 233-252.

HARDER, B.: Die übersinnlichen Phänomene im Test. Augsburg 1996.

HENSEL, W.: Pflanzen in Aktion. Heidelberg 1993.

HRUSHESKY, W. J. M.: Circadian pharmacodinamics of anticancer therapies. In: Clinical Chemistry 39. 1993, 2413-2418.

KELLER, H.-U.: Astrowissen. Stuttgart 1994.

KELLER, H.-U. (Hrsg.): Das Himmelsjahr.

KLEINHOONTE, A.: De Door het licht geregelde autonome bewegingen der Canavalia-bladeren. Delft. Proefschrift Utrecht. Rijksuniversiteit Wis. en Natuurkunde 1928.

KUMM, W.: Gezeitenkunde. Bielefeld 1992.

LEMMER, B.: Chronopharmakologie – Tagesrhythmen und Arzneimittelwirkung. 2. Auflage. Stuttgart 1984.

LEMMER, B.: From the biological clock to chronopharmacology. Stuttgart 1996.

LEWY, A. J.: Effects of light on human meltonin production and the human circadian system. Progress in neuro-psychopharmacology and biological psychiatry 7. 1983, 551-556.

LIEBER, A. L.: Guter Mond, böser Mond. Düsseldorf 1997.

MEISSL, H./ EKSTRÖM, P.: Das Pinealorgan und sein Hormon Melatonin. In: Spektrum der Wissenschaft, Heft 7. 1996.

MILES, L. E. M./ RAYNAL, D. M./ WILSON, M. R.: Blind man living in normal society has circadian rhythms of 24.9 hours. In: Science 198. 1977, 421-423.

MOOG, R./ HILDEBRANDT, G./ PLAMPER, H./ STEFFENS, B.: Circadians rhythms and circadian synchronisation in blind persons. In: MORGAN, E. (Hrsg.): Chronobiology & Chronomedicine. Frankfurt am Main 1990, 52-55.

MOORE, R. Y.: Organisation and function of a central nervous system circadian oscillator. In: The suprachiasmatic hypothalamic nucleus. Federation proceedings 42. 1983, 2783-2788.

MLETZKO, H.-G.: Biorhythmik. Wittenberg 1985.

ORLOCK, C.: Die innere Uhr. Stuttgart 1995.

OTT, J.: Meereskunde. Stuttgart 1996.

PALMER, J. D.: The biological rhythms and clocks of intertidal animals. New York 1995.

PRESTON, F.: Further sleep problems in airline pilots on world-wide scedules. Aerosp. Med. 44. 1973, 775-782.

REINBERG, A./ SMOLENSKY, M. H.: Biological rhythms and medicine. Cellular, metabolic, physiopathologic, and pharmacologic aspects. Berlin 1983.

RENSING, L.: Der molekulare Mechanismus der circadianen Uhr. In: Biologie in unserer Zeit, Ausgabe 25. Weinheim 1995.

RENSING, L.: Biologische Rhythmen und Regulation. Stuttgart 1973.

RODEN, M./ KOLLER, M./ PIRICH, K./ VIERHAPPER, H./ WALD-HAUSER, F.: The circadian melatonin and cortisol secretion pattern in permanent night shift workers. American Journal of Physiology 64. 1993, R261-R265.

ROSATO, E./ PICCIN, A./ KYRIACOU, C. P.: Circadian rhythms: from behaviour to molecules. In: Bioessays 19. 1997, 1075-1082.

SCHÄFER, H.: Astronomische Probleme und ihre physikalischen Grundlagen. Braunschweig 1988.

SIMONS, P.: Pflanzen in Bewegung. Basel 1994.

SPIE? H.: Chronobiologische Untersuchungen mit besonderer Berücksichtigung lunarer Rhythmen im biologisch-dynamischen Pflanzenbau. Band 3. Darmstadt 1994.

SUNDARARAJAN, K.S.: Environmental control of circadian rhythms in plants. Bikaner 1986.

VON DER HEYDE, F./ WILKENS, A./ RENSING, L.: The effects of temperature on the circadian rhythms of flashing and glow in Gonyaulax polydera: are the two rhythms controlled by two oscillators? In: Biological Rhythms 7. 1992, 115-123.

WATERMANN, T. H.: Der innere Kompaß. Heidelberg 1990.

Index